A NOTE ON THE AUTHOR

Timandra Harkness is a writer, comedian and broadcaster, who has been performing on scientific, mathematical and statistical topics since the latter days of the 20th Century. She has written about travel for the *Sunday Times*, motoring for the *Telegraph*, science & technology for *WIRED*, *BBC Focus Magazine* and *Men's Health Magazine*, and on being 'Seduced by Stats' for *Significance* (the Journal of the Royal Statistical Society). She is a regular on BBC Radio, resident reporter on social psychology series *The Human Zoo*, and writes and presents documentaries and BBC Radio 4's *Future Proofing* series.

In 2010 she co-wrote and performed *Your Days Are Numbered: The Maths of Death*, with stand-up mathematician Matt Parker, which was a sell-out hit at the Edinburgh Fringe before touring the rest of the UK and Australia. Science comedy since then includes solo show *Brainsex*, cabarets and gameshows. Meanwhile, she puts her MC skills to more serious uses, hosting and chairing events with Cheltenham Science Festival, the British Council, the Institute of Ideas, the Wellcome Collection and a Robotics conference in Moscow, among many others.

Also available in the Bloomsbury Sigma series:

BIG DATA

DOES SIZE MATTER?

Timandra Harkness

BLOOMSBURY
sigma

Bloomsbury Sigma
An imprint of Bloomsbury Publishing Plc

50 Bedford Square
London
WC1B 3DP
UK

1385 Broadway
New York
NY 10018
USA

www.bloomsbury.com

First published 2016. This paperback edition 2017.

British Library Cataloguing-in-Publication Data
A catalogue record for this book is available from the British Library.

Every effort has been made to trace or contact all copyright holders. The publishers
would be pleased to rectify any errors or omissions brought to their attention
at the earliest opportunity.

Extract on pp. 85–86 from A. M. Turing, 'Computing Machinery and Intelligence',
Mind, 1950, LIX, 236, by permission of Oxford University Press.

Library of Congress Cataloguing-in-Publication data has been applied for.

ISBN (paperback) 978-1-4729-2007-2
ISBN (ebook) 978-1-4729-2006-5

2 4 6 8 10 9 7 5 3 1

Illustrations by Timandra Harkness

Typeset by Deanta Global Publishing Services, Chennai, India
Printed and bound in Great Britain by CPI Group (UK) Ltd,
Croydon CR0 4YY

To find out more about our authors and books visit www.bloomsbury.com.
Here you will find extracts, author interviews, details of forthcoming events
and the option to sign up for our newsletters.

For Linda,
who would have made this a better book,
were you here to read it.

Contents

PART 1: WHAT IS IT? WHERE DID IT COME FROM?

'What is this book?' asked my stepmother, Juliet.
'Is it for people like me, who keep hearing the phrase 'big data' and want to be able to talk about it at dinner parties?'

'Yes,' I said, 'that's *exactly* what it is.'

Not only for Juliet, and not just at dinner parties – it's a book for anyone who gets the feeling big data is interesting and important, and should be talked about, but doesn't want to study mathematics or computer programming.

In 10 chapters I aim to get you from the most basic ideas to some of the thorniest issues we need to be arguing about.

On the way, you'll meet some of the people, ideas and projects I've been lucky enough to encounter around the world. Much of this book is in other people's words, telling their own stories or introducing concepts that help me understand why big data matters. I've tried to structure it so each new idea builds naturally on what's gone before.

That means it's written to be read in order. You can dip in and out if you prefer, of course. Hey, it's your book, you can wallpaper your bathroom with it if you like.* But I think you'll get more out of it if you read from beginning to end.

Big data is a huge subject, and changing so fast I sometimes felt I was running as fast as I could just to stand still. The extra chapter I've added for the paperback edition is probably out of date already. The subject matter of any one of these chapters could fill an entire book. So there are things I only touch upon, or miss out altogether. It doesn't mean they're not important or

---

*Unless you got it out of the library.

interesting. I hope I will give you enough of an overview that you will be able to go and find out more for yourself.

I have my own opinions on what is great, and not so great, about big data. I don't want you to accept them. I want you to make up your own mind. That's kind of the point of the whole book.

But just as important to me is that you enjoy reading it. I hope you do.

CHAPTER ONE

What is data? And what makes it big?

What is data?*

Thirty thousand years ago, in central Europe, somebody scratched 57 notches into a wolf bone. Those 57 notches, grouped into fives just as you might tally something† today, are the earliest known recorded data.

We don't know anything more about who scratched them, or even if the notches were all made by the same person. We have no idea what they denote, only that they were a record of something. Which may not seem like much to you, but it represented a breakthrough in how our ancestors were able to keep track of things.

Imagine for a moment that the fact it's a wolf bone is significant, and knowing how many wolves you'd killed was important for some reason. Perhaps you wanted to see if the local wolf pack was getting bigger or smaller, or whether the new flint-tipped arrows were more efficient than the old wooden ones, or just to win an argument about which member of your tribe was the best wolf-killer and got to sit nearest the fire.

---

* If you're thinking, 'what *are* data! The word 'data' is the plural form of *'datum'*, the Latin word for something given!' then this may not be the book for you. Personally, I regard data as rather like butter: stuff that can be created in lumps of different sizes, and used to enhance our lives. More cynically, you could spread it thickly to make something more appetising. Nobody says 'butter are …'

† I'm keeping a tally of how many cups of tea I'll drink as I write this book. The publisher reckons 550, which I think is risibly low.

You could hang on to a trophy from each wolf, and just see which pile of skulls is biggest, but that takes up room, and is vulnerable to being eaten by dogs. If you can represent each wolf with a notch, all you have to do is compare bones and see which has more notches.

Somebody in an Ice Age cave, in what is now the Czech Republic, had invented digital data.

Today, you can download Wild Wolf Data from the comfort of your own computer. The International Wolf Center in Minnesota, USA, fits wild wolves with tracking collars: radio collars since 1968, and more recently GPS collars that use satellite links to track the wolf's position. This has allowed them to locate individual wolves at any given time, but also to study patterns of wolf movement and behaviour, and even to predict likely conflicts between the wolves and their human neighbours.

The technology is more advanced, but the basic principle is the same: turn your information into numbers, and record it in a form that's easy to use and share. GPS data, tracking wolves through the forests of America, is digital information, by which we simply mean that it comes in numbers that you could, in theory, count on your fingers, your digits.

You'd need a *lot* of fingers, but that's where computers come in handy.

Today, computing technology is so cheap, compact and powerful that domestic washing machines use computers to control laundry cycles. Impressive. And yet it's still easier to keep track of wild wolves in Minnesota than of your own socks.

Without computers, big data would be impossible, so let's take a quick look at their unstoppable rise.

A century of computers

The earliest computers wore petticoats. Until the twentieth century, 'computer' was a job title, and people, mainly women, were paid to do mathematics, with the aid of primitive technology such as log tables and slide rules, both of which were still in use well into the space age.

The first computer in the modern sense was built by IBM in 1944, in partnership with Harvard University. The Automatic Sequence Controlled Calculator, affectionately known as the Mark 1, was 2.4m (8ft) high and more than 15m (50ft) long. It weighed nearly 4,535kg (5 tons) and worked by a combination of electrical and mechanical parts, relay switches, rods and wheels. Computer historian John Kopplin described it as sounding 'like a roomful of ladies knitting'.

The Mark 1 could add together 23-digit numbers in under a second. Multiplication took around five seconds, and division over 10 seconds. It received its program and data in the form of holes punched into paper tape and cards.

Mathematician Grace Hopper became the chief programmer. Her work was central to the development of computer programming, but you may be more entertained to learn that she was the first person to debug a computer: she removed a moth that got stuck in the mechanism.* Most of the Mark 1's early tasks related to the Second World War. Grace Hopper was officially part of the US Naval Reserve, and remained so until she retired, aged 79, with the rank of Rear Admiral (lower half†).

For tasks such as predicting the path of artillery shells, you put numbers in, and you got numbers out. But after the war, both business and government wanted to use the computer for a wider range of tasks. Human beings don't naturally converse in a string of ones and zeroes.‡ They wanted to use recognisable words and syntax to set tasks for the Mark 1, and to understand

---

* A mechanical fault was already called a 'bug', so she appreciated the joke.

† Which makes her sound either like part of a pantomime horse, or the victim of a conjuring trick that went wrong.

‡ Computers work in binary, which is simply a way of expressing numbers using only 1 and 0. When your brain is made of on/off switches, this has advantages. (Cont. p.14)

the answers that came back. Hopper led a team who developed a new programming language* using words and structures from the English language, so that non-specialists could work more easily with computers.

The development of what today we'd call software was a step towards introducing the power of computing into non-mathematical areas of human life. But the hardware was unwieldy and expensive. When Harvard's Howard Aiken, the inventor of the Mark 1, was asked in 1947 to estimate how many computers the US might buy, he said six. It would take a transformation in how they worked, and how they were built, to get us to the present day.

The laptop on which I'm writing this book is smaller and lighter than the machines used to punch the data cards for the Mark 1, works millions of times faster and costs a fraction of the price. Instead of rods and wheels, it uses electronic circuits printed on tiny slivers of silicon: cheaper and less susceptible to moths. The Mark 1 was handmade by experts, but my computer is mass-produced by machines and assembled by people with a few specific skills.

The miniaturisation of electronic components, combined with processes that make them cheaper to produce, gave rise to Moore's Law, coined by Gordon Moore, co-founder of microchip company Intel. Moore's Law says that the amount of processing power you can fit on to a chip will double every

---

You and I can express any number, no matter how large, using only the digits 0–9, by adding columns for tens, hundreds, and so on. After 99, we know it's time to put a 1 in the hundreds column and reset the tens and units columns to 0, giving us 100.

Computers have to add a new column, the Twos column, after they get to 1. Two in binary is 10. Ten is written 1010 (one eight, no fours, one two and no units). So the number of columns goes up very quickly indeed. If a computer offers you a six-figure salary, beware – that's less than £64 a year.

* COBOL, COmmon Business-Oriented Language.

couple of years, while costs of production fall.* In 1965, he predicted that:

> *Integrated circuits will lead to such wonders as home computers – or at least terminals connected to a central computer – automatic controls for automobiles, and personal portable communications equipment. The electronic wristwatch needs only a display to be feasible today.*

You can now carry in your pocket a computer far more powerful than all the computers that existed in the world 50 years ago. The fact that we use so much of this technological power to play games, or count our own footsteps, is an indication of how ubiquitous, how effortless, it has become.

So data can be any kind of information, so long as it's expressed as numbers, in a digital form that computers can store, process and manipulate.

OK, so what's special about *big* data?

At a conference in New York I have coffee with Roger Magoulas, the man reputed to have invented the term 'big data'.

He's diffident, but he does admit that he first used the phrase in 2006, 'and after that the term started being used a lot more'. In 2009, he contributed to a special big data issue of O'Reilly's *Radar* newsletter, taking examples from Barack Obama's US presidential election campaign and fast-growing social media sites. Magoulas spotted some new developments that went beyond size.

Prediction, for example. Companies weren't just analysing the past, they were using data to look forwards as well as backwards.

---

* The details of the 'law' have changed over the years, but the prediction of exponential growth, repeated doubling, is still widely believed.

And instead of data they collected themselves, people were using the masses of information available on the internet. It might be indirect information, what Magoulas calls 'faint signal' data, but with enough of it and the right techniques, it could give answers.

Those techniques meant harnessing machines that could teach themselves. Machines could learn to make sense of information, even when it came in a form designed for human-to-human communication.

Magoulas is a polymath. He does write computer code, but he's equally interested in the human side, in asking the right questions, in understanding the ways that human beings can draw meaning from digital information.

'There's a few things you can automate,' he says, 'but most of it is to augment people. Nothing should make the decision for you, it should make you a better decision-maker because you're getting these new inputs.'

Big data is sometimes described in terms of three Vs, defined by an analyst called Doug Laney in 2001 when it was plain 'data'. Volume, velocity and variety identify three of the qualities that Roger Magoulas also noted: there's a lot of data, it's coming at you very fast and in different forms.

But I have my own acronym[*] to sum up what's special enough about big data to be worth writing (or reading) this book: big DATA.

Big is for big, obviously. DATA spells out four key elements that make it new and distinctive: it deals with many Dimensions, it's Automatic, it's Timely, and it uses AI, Artificial Intelligence. I'll go through those one at a time:

Big

It's difficult to define the bigness of data in absolute terms. Partly because it's expanding so fast that between me typing

---

[*] Strictly a back-ronym, because it's not a coincidence that I end up with the words I'm trying to define.

a number and this book being printed, it would already be out of date. To give you an idea, O'Reilly's 2009 big data special reports scientists handling 'some of the largest known data sets' of several petabytes.

A petabyte is 1,024 terabytes. On my desk is a portable hard drive that fits into my pocket. I used it to back up the manuscript of this book, and pretty much everything on my computer, plus my entire music collection, but it's far from full. It holds 1 terabyte, 1TB, of data, cost me less than US$100, and could contain nearly 200,000 copies of the complete works of Shakespeare.

I could fit 1,024 of them, a petabyte, into a large suitcase. So what would have been one of the largest known datasets in 2009 would now fit on to a luggage trolley.

There are very few measures that make sense on an everyday level. In the internationally recognised unit of measurement for Very Large Things, the 'to the Moon and back', if all the information currently available as digital data could be put on to CDs, it would stretch to the Moon and back between three and 20 times. Though, by the time you read this, that'll be 100 times, or 1,000.

Does that help? CDs are already old-fashioned in computing, because they don't really hold enough data. One reason the world's stock of data is growing so fast, doubling every three years by some estimates, is that we use more data to say the same thing. If you have a camera in your cellphone, it's probably 10mp or more – that's 10 megapixels, 10 million cells of colour and light, in that photo you took of your mates in that bar. No wonder it takes nearly 3MB of data to store it.

So part of the proliferation of data is deceptive; we're just recording the same things in more detail. But there is genuinely more of the stuff. Filing cabinets full of paper have become computer servers full of digital data, which is more compact to store, and easier to find. So what?

Data analysts talk about 'data mining', as if all the information is already buried beneath our feet, and we just need to dig down through the dirt to bring back the diamonds. Big data has an air of completeness, of everything already being in there somewhere. Instead of asking questions with a survey, data analysts put queries to the data that's already collected. This is a big change in how information is understood.

If you read that 93 per cent of women agree a certain face cream is brilliant, you may be impressed. If, like me, you check the small print and find it was 93 per cent of a survey of 28 people, commissioned by the manufacturer, not so much. But if the same company had somehow surveyed all 331,548 women who bought the product in a year, and 93 per cent of them say it's brilliant, the face cream may be worth a try.

It doesn't take a degree in statistics[*] to understand that a selective sample isn't a completely accurate guide to the whole picture.

Scientists use the letter n to tell you how many items they studied: 'n = 11' means you had 11 wolves, or patients, or women using face cream, in your study or experiment. Now data scientists talk about 'n = all', meaning they have the whole population in their dataset.

Somebody, somewhere, still had to decide which information to collect, but it's easy to gather that data just in case, and decide later whether it's useful.

D is for dimensions

A space scientist, Dr Sima Adhiya, once told me a story about her grandmother. In India, where the grandmother lived, the crickets were a constant background noise. And she told her granddaughter that the song of the crickets could tell you how hot the weather was that day.

When Sima grew up and became a scientist, she discovered that her grandmother was right.

A scientific paper, *The Cricket as a Thermometer*, published in 1897 by Amos E. Dolbear, expressed the relationship

---

[*] Which is lucky, because I don't have a degree in statistics as I type this. If things go well I should have one by the time the paperback goes on sale.

between temperature and how fast the crickets chirp in an equation known as Dolbear's Law.[*] So if you didn't have a thermometer, but you were within earshot of some crickets, you could tell the temperature to the nearest degree by counting chirps with a stopwatch.

How did Dolbear discover this? Tantalisingly, he doesn't say. His main interest was in turning sound waves into electrical signals, and vice versa. He invented something very like the telephone before Alexander Graham Bell, and patented the wireless telegraph before Marconi. So it's possible that he used some ingenious apparatus to turn the cricket sounds into electrical waves before measuring their frequency.

But making a note of the temperature on successive days, and counting chirps per minute at the same time, would be enough for him to find the mathematical relationship. Temperature and chirps per minute are two very different types of thing, but by expressing both as numbers, and treating them as different dimensions of the same moment in time, Dolbear found a correlation close enough that one could predict the other.

Anything that can be turned into numbers can be a dataset. I could compare my tea-drinking against words written every day, and turn it into an equation to predict how many teabags I will need to finish this book.[†] I could go further and download weather data, to see if the weather has any effect on how much I write and/or how much tea I drink.[‡]

---

[*] Expressing temperature as T degrees Fahrenheit, and chirps as N per minute, Dolbear's Law tells us that $T = 50 + (N - 40)/4$. To find temperature in Fahrenheit, count chirps per minute, subtract 40, divide by 4, then add 50.

[†] Knowing my capacity for procrastination, I'm surprised I haven't.

[‡] I'm assuming that neither words written or tea imbibed will have any effect on the weather, though I suppose boiling the kettle might add a little to both the local temperature and the water vapour in the atmosphere. So I might be very slightly increasing the chirp rate of the local cricket population with each cup I drink.

If Dolbear were alive today, he could use a digital recording device to record the song of the crickets, and a computer to analyse the frequencies and compare them to the readings from a digital thermometer. In fact, he could write a computer program to do it all for him. Although, as he'd be nearly 180 years old, he might prefer to hire some young person to write it for him. Then he could get back to squinting at the controls of his computerised washing machine.

Having data in digital form is the first step towards making this kind of pattern-spotting possible, and it means you can link datasets of very different types. Perhaps D should stand for Datasets instead of Dimensions. Or for Diverse. The point is that you can now combine utterly Different types of information to learn something new.

A is for automatic

Think of how many things you do every day that involve computer technology.

In London, where I live, you can no longer pay cash to travel by bus. You can use an Oyster card, or you can tap a bank card directly on to the yellow pad on the bus. Whichever you use, the travel company deducts money from your account. When I run out of credit on my Oyster card, as I regularly do, I used to have to pay a higher cash fare to get a paper ticket. Now I just swipe my debit card.

The ease of collecting data by machine makes it very simple to gather up more than just a total of fares taken. I haven't registered my Oyster card with my name or address on the Transport for London system, which is why it runs out, instead of automatically buying itself extra credit when the balance falls below £10. I do, however, top it up using my bank or credit card, so it's already linked in their system with my name and address.

The more we do everyday things via computers, cellphones and plastic cards, the more information is automatically hoovered up and stored on a computer somewhere. Transport for London doesn't only know how many passengers boarded their trains, buses and trams, but also where we got on and off[*] and a swathe of other information such as where I live, and possibly whether this is my regular commute.

When information was collected by people, who had to write it down or type it into a machine, decisions about what to collect were tough. Now we're far beyond recording data being easier than using flint to make notches in a wolf bone. In many cases, it's now easier to record data than *not* to record it. Recording data is the default.

Every time you use a cellphone it stores all sorts of information in digital form: not only the numbers you call, and how long you talk for, but where you are whenever your phone is turned on. If you have a smartphone, it's full of cunning little bits of kit, such as accelerometers and GPS receivers. That's how the wonderful apps work that let you point your phone at a star and find out it's the planet Jupiter.

You may already be one of the people using technology to capture your own data. Many cellphones come preloaded with apps related to health and fitness. You can track how many steps you take, how many calories you burn, even your heart rate. Or go further and turn other aspects of your life into data. How happy are you? How many tweets have you sent? How many cups of tea have you drunk today?[†]

It's easy to automate both the collection of the data and the processes that turn it into useful information. In order to keep a tally of how many steps you take every day, your smartphone performs detailed calculations using the accelerometers that track changes of angle as you move. You don't want to read those calculations. You just want a total of steps that day.

Or do you? Perhaps you want to combine it with other information. If you wanted to lose weight, you could combine it with an app that tells you how many calories you're taking in by analysing photographs of your meals. If you're

---

[*] Apart from the buses, which just record when you get on.

[†] Three so far.

competitive, you could share your total online and compare yourself against others. You might want a map of where you went, like Murphy Mack from San Francisco.

Murphy, like many keen cyclists, uses an app called Strava. Using GPS technology in a device such as a smartphone or satnav, Strava tracks your route, your speed, and how far up and down hill you went. Then it turns that data into various formats, such as a training calendar, personal best records, and a map of where you went.

Murphy rode his bicycle for 29km (18 miles) through San Francisco and uploaded his data. It wasn't just his own pulse racing after that ride, because the red line on the map drew a heart shape. Not the most perfectly rounded heart shape, as San Francisco's street map is mostly a rectangular grid, but discernibly a heart, with 'marry me, Emily' spelled out inside.

If you get marks for effort when proposing marriage, Murphy gets top marks. San Francisco is far more hilly than the neat map grid suggests. Small wonder that his 80-minute ride burned off 749 calories, so we can tell exactly how much effort he went to, physically at least. And Emily said yes.

This is a frivolous example, though it's very important to Murphy and Emily. But the very fact that you only need a free app, a smartphone and the internet to automatically turn your cycle ride into data, and back again into a romantic picture, reveals how easily big data becomes part of your life.

T is for time

Have you had your first barbecue of the year yet? Or are you waiting for the weather to get just a bit warmer? How far ahead will you decide? A couple of weeks or half an hour beforehand when you're out shopping and see the sausages on display?

We may not need to plan ahead, but supermarkets do. If they get it wrong, they'll either miss out on millions of sales of burgers, charcoal and beer, or be left with unsold stocks of perishable meat and hot dog buns. How can they predict the first big barbecue weekend of the year?

The weather is one factor. Supermarket chain Tesco calculated that a 10° Celsius (18° Fahrenheit) rise in temperature means they will sell three times as much meat. In other words, we all want to barbecue in spring. After June, even though it's warmer, the novelty of charred meat and marauding insects wears off.

Data analytics company Black Swan uses more sources of data, captured almost in real time, to give supermarkets a more reliable prediction. By collating extra information, such as how many people are searching 'BBQ' online or the general mood of social media posts, they claim to have saved at least one supermarket chain millions of dollars in wasted stock or missed sales.

This example shows two characteristics of big data. The picture changes continuously, as data comes in almost as fast as it's produced. And because the software produces not just a snapshot but a moving picture, it can be extended into the future.

You may not care whether you can buy barbecue food when the mood takes you, or whether retailers have to throw away unsold chicken drumsticks. But the same approach can be used to predict more important things, such as medical needs: the spread of colds and flu, for example. Black Swan worked with a large pharmaceutical company that wanted to know which areas were being hit by those seasonal illnesses, or were about to be hit. This was for commercial reasons. They knew which of their products sell better when people have colds, or don't have one yet and are trying not to catch one.

By combining the company's own sales information with weather forecasts, web searches and social media posts, data analysts were able to identify, down to postcode level, trends in the spread of coughs and sneezes. The company was able to provide targeted online advertising in the places most likely to need vitamins and cold remedies.

Black Swan are now applying similar techniques to predicting how many staff need to be on duty in hospital Accident and Emergency departments. Previous hospital records help, of course. But by adding weather forecasts to the mix, social media posts by people saying they're going out for the night, and even mentions that they're #drunk, the system can foresee how busy the night will be for emergency doctors and nurses.

This kind of predictive analytics is in demand, and Black Swan has expanded rapidly. So rapidly that in four years of existence they've had to move office nine times. Which suggests a certain failure of prediction when it comes to their own future needs. Next time, they should try using their own analytics system.

But what is this system, and how does it work?

A is for AI

And AI is for Artificial Intelligence.

Try to write a definition of a cat. Not just a dictionary entry, but a description that would equip an alien who's never seen one to distinguish cats from other creatures, or from non-living things. Tricky, isn't it? Now imagine writing a definition of a cat that would allow a computer to reliably distinguish cat from non-cat, without having to take DNA samples from a cat.

But if I show you 100 pictures, you can easily sort them into cats and non-cats because you've seen a lot of cats, and spent your early years pointing to furry things, saying 'cat!' and having some adult either encourage you or correct you with 'dog!' or 'rabbit!' or 'your uncle Jock's sporran'.

That's the underlying principle of machine learning. A machine teaches itself the difference between cat and non-cat by being shown a lot of pre-sorted pictures, getting some feedback on how well it's doing, and refining its sorting process as it goes along.

There are different types of AI, and some computers combine several types in one process. Some use the kind of machine learning that sorts cats from dogs, and can also learn to recognise faces and spot other patterns. Some of them use pure logic, the kind used in mathematics to construct a proof that $1 + 1 = 2$.[*] Others place bets on uncertainty, drawing useful conclusions from incomplete knowledge, or work with 'natural language', the kind of language a human would use.

Is this thinking? I would say no, because it fundamentally lacks some aspects of human thought. You may not care, so long as it does the task we have set it, but people working in AI are often as interested in these philosophical questions as in the technical aspects.

As Hurricane Joaquin brushes past New York in 2015, I'm in a rooftop bar with Tim Estes.

The work of Austrian philosopher Wittgenstein was Tim's first love, but instead of pursuing a career in academic philosophy, at 20 Tim set up his own AI company, Digital Reasoning. Now, the technology he calls 'cognitive computing' is at work in projects such as online child protection, health care and government intelligence.

It's not a complete change of direction. Wittgenstein was interested in knowledge and meaning, how we know what we know, and the troublesome mismatch between pure logic and the way the real world works. These are the same questions Digital Reasoning tackles on a more practical level. This rain-lashed bar, the decorative trees outside trembling and whipping against dark skies, is Tim's twenty-first century version of the philosopher's wooden hut.

Data is often compared to oil, as the raw material that will power the next industrial revolution. Taking up this analogy, Tim Estes talks of AI as the engine that will put it to use.

When oil was first extracted from the earth, he points out, it was mainly used in kerosene lamps. This was important, in

---

[*] This may seem obvious, but Bertrand Russell and A. N. Whitehead spent 360 pages of *Principia Mathematica* proving it.

an era before electricity, and it replaced whale oil.* But only when the internal combustion engine came along could oil transform civilisation, by powering industry and transport. Still later, the jet engine brought kerosene back into use, shrinking the world through flight. Having the fuel is only the start.

Tim thinks we're still at the propeller aeroplane stage, and that's why we need better artificial intelligence. He does use the words 'reasoning' and 'thinking' for what machine learning does, but his vision is not of a world where the machines take over and we are redundant, but one where computers take on the 'cognitive labour' we don't want to do, in the same way that oil-burning engines released us from manual toil.

The transitions won't always be easy, and there will be losers as well as winners. But even at this propeller-driven stage, there's plenty of imagination taking flight.

So that's the theory of big DATA. To see what it can do in practice, let's buzz over to Southern California and see big data at work on small things that fly, crawl and bite.

Bug data

The September sun is hot enough to drive me into the shade for lunch. I have to push my empty chilli bowl across the table to distract the flies, my own little skirmish in humanity's ongoing war against insects. A plump bluebottle settles on a scrap of meat, long mouth reaching out to plunder my leftovers.

I'm here to talk to a man who's harnessing big data in that war, though as he points out, insects can also be useful to us, when they're pollinating our food crops for example. So some kind of insect genocide would be a very bad idea, even if it were possible.

'On the opposite side, insects maybe eat or destroy about $400 billion worth of food every year,' says Eamonn Keogh,

---

* Yes, fossil fuel helped to save the whale.

who has somehow kept his Dublin accent intact after 30 years in America.

'If the insects agreed not to eat our food we could feed the world five times over, but we constantly have to fight this battle against the insects for our own food', he says. 'That's in agriculture, and on the other side we have the problem that insects spread diseases. Most notably, mosquitoes spread malaria, but there are other insects and diseases: dengue, chikungunya, West Nile fever, there's a whole list of them.'*

Eamonn talks 19 to the dozen, bursting with ideas and enthusiasm. But he's a Professor of Computer Science and Engineering at the University of California Riverside. So, jokes about debugging aside, why is he working on insects?

'I want to treat insects as though they are digital objects,' he says, comparing them to email. 'I have one algorithm and it puts my real email in this pile and my fake email, my spam, in this pile. And because it's digital I can do two things that are very useful.

I can press delete and I can press forward. What I want to be able to do is to delete insects and forward insects.'

I have a mischievous vision of forwarding wasps to somebody else's office, but this is all in the future. For now, Eamonn's main focus is identifying and sorting insects, the same way your email inbox knows which email is from your boss and which is spam.

His lab is a curious mixture of hi- and low-tech. There are sleek computer terminals at which his PhD students, freshly returned from summer placements in tech companies such as Yahoo! and Facebook, work quietly. But there are also plastic boxes full of real live insects, with damp flannels draped across the gauze lids to keep their conditions humid. And a poster of *Mosquitoes of the Midwest*.

There are soldering irons, racks of electronic components, and Lego. 'I'm living the dream, I am,' jokes Eamonn, 'I'm a grown man playing with Lego.' But there's a good reason for it.

---

* Including, as we've all been made aware since I visited Eamonn, the Zika virus.

What they're building here is an ingenious, adaptable, portable insect-sorting device, like a physical spam filter for flying bugs. Using components that snap together means the design can be adapted, expanded and assembled by anybody, anywhere in the world.

In his office, Eamonn shows me the current state of insect surveillance technology, especially in developing countries. It's a piece of yellow cardboard marked out in squares.

'These are classic. They're gluey here, and they may have an attractant. These work reasonably well for agricultural pests.'

The farmer places them in the field, goes back a week later, counts the number of insects stuck to each square, and reports the results. Eamonn lists the drawbacks.

'First of all, it's expensive. These things cost 50 cents, in Africa 50 cents is a significant part of a person's monthly salary. Secondly, it's quite inaccurate because these could be five different species here, 10 different species, but they look like brown dust to us, so you may get the wrong species. And finally there's a time lag. If the farmer goes once a week, some pests only live for two or three days as adults, so by the time you count them they've already done all the damage. A week lag is just incredibly slow.'

Eamonn gives his verdict, 'Current methods of surveying these insects are inaccurate, with a long time lag, and rather expensive. So my idea was: Let's make insect surveillance digital.'

Which is where the Lego and electronic components come in.

Like Dolbear with his cricket thermometer, Eamonn's lab uses the vibrations that insects produce as a source of information. However, instead of using the frequency to tell them something about the temperature, Eamonn wants to know more about the insects themselves. And instead of listening directly to the sound, they're using lasers.

'We use red lasers because insects cannot see them. You can see this red light here, but an insect can't as it flies through. So it doesn't change its behaviour.'

The lasers shine on to photodiodes, electronic components that translate light hitting them into an electrical signal. So anything interrupting the light changes the electrical signal coming out, which is translated into sound, so you can hear what is passing through the light gate.

'If the light is a constant amount of light, you get basically a flat line,' says Eamonn. 'If I put my finger in here I get a blup blup sound. If I have a tuning fork – *ding!* – and I put it through, I hear the tuning fork beautifully. We actually use it to calibrate the sensor.'

So if an insect flies through the gap, you can play back the recorded signal and hear a recognisable insect sound, only without the background noise.

This is the first stage, the collection of digital signals, each corresponding to a specific insect that has flown through Eamonn's laser gate. Once the sensors are built and set up, collecting that data is as automatic as scanning your travel card as you go through the gates at the station. More automatic, in fact, more like scanning your card as you walk through the gate, without you taking it out of your pocket. And you don't even need a card, so it's more like a gate you don't know is there, that recognises your gait, the rhythm of your walk.

Next, to make sense of that data. Using slightly more than the *Mosquitoes of the Midwest* poster.

'In order to understand, and then intervene and change the world, you have to know what the problem is, and this is more complicated than it seems,' says Eamonn.

'People say "mosquitoes", but there are 3,528 kinds of mosquitoes. There's actually one mosquito called the London subway mosquito. It literally evolved, very recently obviously, adapted to the London Underground. It's the only place in the world it's found, and it's a distinct species.'

Eamonn distills the problem, 'To our eyes they're just flying brown things, but the intervention that will work on one won't work on the other one. You have to understand the individual's behaviour, the time that they appear, which chemicals attract them, which chemicals repel them and so forth.

So what's crucial here is surveillance to know what you have, where you have it, when you have it.'

Luckily for Eamonn, one way to distinguish insect species is precisely by measuring the pitch of their whine, even when it's too high-pitched for the human ear. The exact frequency will vary, but just knowing that an insect flew through, beating its wings 600 times per second, might be enough to tell you it belongs to one species of mosquito.

Suppose you have two species with similar frequencies, but one tends to come out at night and the other doesn't? Then the time or location of the observation could be a deciding factor. So having a number of different dimensions for one observation really helps.

And, because this is big data, we're not relying on Liudmila, Nurjahan and the other PhD students to waste their considerable brain power on analysing this information. Instead, they have a computer using deep learning, a form of artificial intelligence that has worked out for itself the best ways to sort the incoming signals, by comparing known insects, identified by hand,[*] with their acoustic signatures.[†]

This kind of AI is one of the key aspects of big data. And, for Eamonn, the sheer quantity of data is vital to that process. With only a few observations of insects, the software can't refine its system, and can end up sorting wrongly, or finding patterns where none exist.

'In the entire world, before I started this, you had like 1,000 data points for all the insects they could look at. Within a few years I had 10 million data points. Just that one difference, nothing else, accounts for almost all the improvement I've made. Just getting lots and lots of data.'

For Eamonn, at least, size really does matter.

---

[*] Yes, that probably was some poor student's job.
[†] Strictly, pseudo-acoustic because they were collected using light, not sound.

'I don't have to build a model of how an insect flies. Taking this agnostic idea and building essentially no models but lots of data got me a large fraction of this way.'

This is a distinction that's often made by people using big data. Instead of starting with a theory and testing it against what you observe, you just collect enough data and see what patterns come out. You still don't know why this particular insect feeds half an hour before sunset, but you do know enough to say that staying indoors for that hour will reduce your chance of getting bitten by 90 per cent.

Using this deep learning approach, a computer that teaches itself to classify insects, combined with masses of observations with many dimensions, is achieving some impressive results. Distinguishing between 3,528 species would be impressive enough, but Eamonn's team can also tell you which sex your insect was, and whether it has already had a meal of blood. Both of which may be important.

Malaria is a big international health problem. If you're George Clooney you can do some work in Darfur, catch malaria and have 'a bad 10 days', as he put it himself. If you're an underfed child already fighting off several infectious illnesses, it can kill you. And malaria is passed around by mosquitoes, who take blood from an infected person and pass on malaria parasites to a new person.

Eamonn has a lot of respect for his mosquito adversaries. He makes them sound like tiny flying ninjas, or the best secret agents in the world.

'She can smell you from at least 200m away, just by your exhaling carbon dioxide. And of course you can never stop exhaling carbon dioxide until you die. So she will find you, in the dark, in the rain, each raindrop is 50 times her mass. Doesn't care. She lands on you. She knows that if she pricks you, you'll kill her.'

Eamonn slaps his arm in illustration. 'So she puts in this chemical, like a dentist rubs on your gums before he puts the needles in, a numbing agent. She sticks in her little stylet, slicing as she goes in, pulls out your blood. She is basically a flying surgeon. She'll triple or quadruple her mass and then just fly away, which is an impressive feat by itself, right?'

And Eamonn does mean 'she', because it's the female mosquitoes that suck the blood. An all-male mosquito population would not spread malaria. Though they also wouldn't survive very long as a population.

Good reasons for knowing whether a mosquito is male or female.

The Sterile Insect Technique, SIT, stops successful breeding by releasing lots of sterile males into the wild. They don't know they're sterile, and neither do the females, so they breed as normal, but have no offspring. So next year, the insect population drops, and may even die out over a few years. All this, without any use of insecticides.

'You could do something like this with mosquitoes,' says Eamonn, 'but the problem is, you can't sterilise millions of mosquitoes and release them all because the females can still bite and give the disease to you.'

You need to release only the sterile males into the wild. They won't bite, but they will mate with the wild females, without fathering any baby mosquitoes. That means fewer mosquitoes to bite you next year.

'So I need a magic room where I raise all the mosquitoes, and the boys can leave as fast as they want, but the females can never leave.'

Like a Hotel California for bloodsucking insects. People have tried this before, but nobody's built a 'magic exit' that works. Eamonn's plan uses another laser, this time powerful enough to zap the females when they try to fly out.

'If I can get this to work, a few factories all about Africa where you grow mosquitoes in large numbers and the males keep leaving …' or even, where you box up the males and send them out for delivery by drone to villages all over the region. Like Amazon, but for sterile male mosquitoes.

'With the sterile males flying everywhere, mating these females, the population would crash. Even eradication is possible, at least on islands or in small places.'

So equipment that sorts males from females could make a life-saving difference to large parts of the world.

Eamonn's big data approach demonstrates the potential of using big sets of data, measuring different *dimensions*, using *automatic* sensors that record and classify their subjects, getting results in a much shorter *time*, and demonstrating how *AI* can make a difference. It's a beautiful case study of big DATA.

And the collection and classification of data are barely beginning. That's what the portable insect traps are all about.

'We're going to build these sensors, these nice little kits, and anybody who wants one can get one. We hope to send these to schools all over the world. You might be in a school in Oxford or in Zimbabwe, or in Adelaide Australia, you take this kit into a high school, capture a local insect or two … we give them a free sensor and they give us the data.'

That's why it's so important to make the detectors cheap, light and very easy to assemble. And when they've returned Eamonn's data, they can use the kit for their own experiments.

'They can say: Do flies like pink rather than blue? And put a pink card here and a blue card here and see which way they go. They can use the sensor for their own science, their own fun, but we as a by-product get a lot of extra data.

So we're going to crowdsource this thing all over the world, from professional scientists to elementary schools in Dublin hopefully.'

And, using cellphone technology, some of the kits would not even need humans to relay the results back to Eamonn's data bank.

'My ambition with these sensors is when they are in the field they'll be so cheap you can just drop them and forget them. You'll never go back, you never have to attend them, never actually do anything with them.'

So even sending the data back to the database would happen automatically?

'The sensor itself would record insects and store the information and then, probably at night time when the bandwidth is cheap or free, it'll send these tiny bursts. The information you need to send is very small because it's only a count of insects so it's like a one second phone call basically.'

So the stuff in the sensor will analyse what the insects are?

'Yes.'

And will basically just send you a little census.

'Exactly.'

Today we captured … and I suppose it can even tell you what time of day?

'Which is important information to know. The only exception would be that occasionally it'll see an insect it doesn't know, and then it would be good to send the entire sound file back to us, and we can listen to it and say: What is that? and do some extra analysis.'

By the time this book goes on sale in California, Eamonn hopes to be getting data from all over the world, covering, 'not every insect, but at least 1 million insects. But we really only care about the 10,000 troublemakers that cause problems for crops and for humans and so forth.

It's hard, and a few years ago I would have thought it was untenable, but I think in a year or two we really could model all the insects that matter. This could sit somewhere on a server here, or in Google, or in the World Health Organization.'

So there would be a global resource detailing where the troublemaking insects are, and what they're up to.

'And now the rest of the world could, hopefully for free, have these sensors in the field, send us questions and we answer back and say: What you have is an Aegypti female who's had a blood meal two hours ago.

That's the vision, the goal.'

That sounds like an ambitious, twenty-first-century project, doesn't it? A global database of insects, and a global network of cheap, portable sensors that can identify them.

Well, try this one for size.

It's called *Premonition*. It involves 12 universities, Microsoft and the US intelligence research agency, IARPA. It's a new way to monitor diseases in developing countries.

'There's a free sensor that senses blood, called a mosquito.' says Eamonn, 'And it's a busy little insect, it will sample all you like for as long as you want. So can we exploit that?'

He answers his own question, 'The global vision is bizarrely ambitious but wonderful. Enormous flying airships, like Hindenburgs, which have hanging from them drones, which have hanging from them insect traps. These drones will leave the mothership, fly non-supervised by humans, and figure out where they can safely leave a trap hanging from a tree or on the ground. Once they leave that, they will return to the mothership and wait.

So now you have these traps on the ground, and these traps will capture mosquitoes that will be returned by drone back to the mothership.'

However, early tests found that the traps were indiscriminate. So when the drones brought them back to the airship, they were full of the wrong kind of mosquito, or the wrong sex, or the right type of female that hadn't had dinner yet. Which must have been the worst possible thing to have loose in your laboratory.

This is where Eamonn's light-gate sensors come in.

'The traps wait till they fly past.' Eamonn points to each imaginary insect flying by. 'No ... no ... no ... yes! ... and snatch it out of the sky.'

He grabs the imaginary mosquito.

'And the drone comes, returns it to the mothership. At the mothership, or on the ground soon after, the contents of the blood meal will be examined, sequenced for all pathogens. So you can say: We think this mosquito bit a chicken, and the chicken has avian fever.'

You don't know which chicken, or exactly where it was, but you do have an early warning system that is a lot cheaper than sending out humans with hypodermic needles to randomly sample local animal or human blood.

'It makes sense because for many diseases if you can intervene very early by quarantine or drugs, the problem can be nipped in the bud for $1 million. A week later that problem could be suddenly a billion-dollar problem. So early detection makes a lot of sense.

No one has ever done a project like this before, and a million scientists will come in and take this data and do cool things that we haven't thought about. That seems inevitable.'

So there we have the last two pieces of the big data kit. You use the data to predict the future, and thereby potentially change it. And you foresee that the data you collect today will be used tomorrow in ways that nobody can foresee.

I wake in the night to the distinctive whine of a mosquito. I can't tell its species, sex or dining status, but I know I left the balcony door open on to the warm Californian night, and that I'm pumping out CO_2 like a neon diner sign for bloodsucking insects. Attempting to hide under the duvet, I wonder whether Riverside's insects carry any diseases I should worry about.

Peaceful sleep is over. In the end, I turn on the light. There it is, not one of the 4mm babies but a massive Flying Fortress of a mosquito, looking for its chance to get at my blood. Without the aid of lasers, I snatch it out of the air. It lies on the bedsheet, legs still twitching.

Doesn't look full, I think. Must have caught it in time.

Only on the flight back to New York do the five bites announce themselves with itching, red bumps. Curse you, mosquitoes, and your sneaky numbing-chemical ways.

Luckily for me, the only thing that infected me during my visit to Riverside was Eamonn Keogh's enthusiasm for big data and its potential. But before we go on to look at some of the many ways in which big data is already being used to change our lives, we need to take a step back in time.

Because before we had big data, we had data, scratched or written by hand. And before we had artificial intelligence we had human intelligence, working out what information to collect, and how to make sense of it.

So I want you to meet some of the people who first collected data, and used it to understand the world. Partly because it'll help when we look at how computers do similar things now. And partly because I'd like you to know their stories, and appreciate how their work still helps us today.

CHAPTER TWO

Death and taxes.
And babies.

Ever since the earliest civilisations, rulers have known that you can't tax something without knowing how much of it there is. Among the earliest surviving written records, pressed into wet clay 4,000 years ago in cuneiform symbols, are Mesopotamian tax receipts.

Even before we could write, our ancestors used clay tokens to record the quantities of oxen, grain and whatever else could be traded and taxed. For all we know, that wolf bone with 57 notches in groups of five was a prehistoric tax return, showing that 57 mammoths had been eaten, and somebody was due to pay their share of cave upkeep, spear maintenance and dung removal charges.

But it's much easier to send a tax demand when you also have a written record of somebody's name and where they live.

No pig left behind

In 1085, King William I of England sent researchers out to record the assets of his kingdom. Another war against Denmark looming, William, better known as William the Conqueror, wanted to know how much tax he could raise, and what military resources he could draft. He had no standing army, so men, horses and weaponry from all over the country were a tax-in-kind as vital as hard cash.

The inspection was thorough. The Anglo-Saxon Chronicle recounts that not a yard of land, 'nor one ox nor one cow nor one pig' was left out. But if you know the King's inspectors are coming to count and tax your livestock, you have a strong incentive to hide them. There may not have been any accountants, or schemes to put your money in an offshore bank account on a Caribbean island,[*] but if you could hide a pig somewhere, or pretend it belonged to somebody else, you might not have to pay tax on it. Conversely, if you could convince the inspectors that a piece of land belonged to you, you could claim it as yours forever.

So a complete census was very labour intensive. Each draft report was checked at a special sitting of the county court, where a jury was asked to verify the area of land, the number of fisheries, plough teams and so on. Only then were the local reports written up neatly in Latin,[†] and collected up to be transcribed into the Domesday Book.

Or strictly, Little Domesday covering the eastern counties of Essex, Norfolk and Suffolk, and Great Domesday covering most of the other counties, though some of them are a bit sketchy because it was never fully completed. The bound volumes, written on parchment with quill pens, leave out the pig-by-pig detail of the original reports and would already have been out of date by the time the last scribe put down his quill and called it a day, after William's death in 1087.

No wonder nobody attempted another national UK census for more than 700 years.

---

[*] Nobody in England was aware that Caribbean Islands even existed in the eleventh century, let alone that they could avoid tax by having a bank account on one. Banks were pretty rudimentary, come to that.

[†] Latin was the most universal language of the day, especially important when 'English' people spoke Anglo-Saxon or French. Just as English today is the international language of science, computing and tourist menus, in the Middle Ages being able to read and converse in Latin meant you could communicate with educated people and order gruel in overpriced inns across the known world. Known to Europe, that is.

Making use of the information in Domesday was hard work, too. Before printing, a postal service or telephones, the only way to retrieve the data in the Domesday Book was to go to where the book was and read it yourself. That meant travelling for days on foot or horseback, along muddy tracks beset by snow, flooding and robbers, to read the one original edition. Assuming you could read, which most of William's subjects couldn't.

Yes, take a moment to be thankful you live in the era of the internet, with all the world's information a few clicks away from wherever you are right now.

OK, that's long enough.

No, you can't just watch the end of that amusing cat video. I want to talk to you about death.

Births, deaths and marriages

Like most people in seventeenth century England, John Graunt was a man of faith. In his case, the Roman Catholic faith, which in those days was politically risky. Nevertheless, somehow he survived both being on the unpopular side of the religious divide and fighting on the losing Royalist side in the English Civil War. So when he buried his young daughter Frances, he must have asked: Why?

He'd already buried his parents in the last 12 months, which was bad enough, but in the natural order of things. Why, when he had lived through wars, political upheavals and repeated outbreaks of plague, had his daughter succumbed to consumption? Perhaps this was part of his motivation for poring over hundreds of death records in search of order and meaning. Or perhaps he just wanted a time-consuming hobby to take his mind off the grief.

The first organised death records in the UK were kept in London when the city was ravaged by plague in the sixteenth century. Individual deaths were already recorded locally, in each church parish, but for the first time that information was collected together in bills of mortality, which also noted the age and cause of death.

The bills of mortality counted deaths in each parish, but not the names of the deceased, making them an early example of anonymous data.[*] As well as providing vital information to the city authorities week by week, the bills were made available to individual citizens, for a price, so they could decide whether to stay in the city or flee the latest outbreak of plague.[†]

As a haberdasher in the City of London, John Graunt would have been active in civic life through the guild system, and wouldn't have found it too hard to get to the bills, which went back over 50 years.

Looking for patterns in the hundreds of lists, Graunt combined the figures into his own tables, organised by age and cause of death. In 1662, he published *Natural and Political Observations Mentioned in a Following Index and Made upon the Bills of Mortality*, which held the record for Least Snappy Book Title until 1929, when Henryk Grossmann published *The Law of Accumulation and Breakdown of the Capitalist System, being also a theory of crises.*[‡]

By putting together a lot of data, Graunt spotted a number of things that nobody had noticed before. He saw that some diseases tended to kill roughly the same number of people every year and others, including plague, come and go. Even accidental deaths such as drowning happened at a stable rate every year. Not exactly the same, but varying around a fairly

---

[*] Though, as some parishes only had one death in a given week, it would have been easy to find out who it was by comparing the bill of mortality with the parish register of births, marriages and deaths. So it's also an early example of data that looks more anonymous than it really is.

[†] And an early example of data as a valuable commodity.

[‡] First published in German as *Das Akkumulations-Zusammenbruchgesetz des kapitalistichen Systems (Zugleich eine Krisentheorie)* – arguably even less catchy.

steady number. In fact, Graunt was probably the first to identify this stable number by adding together the total and dividing it by the number of years, what today we'd call the average.[*]

Graunt set down the horrifying rate at which young children were dying, one in three dead by the age of five. He was also the first to note that more boys than girls were born, a ratio of 14 to 13. And he confirmed that plague was a frequent visitor to the city, not a permanent resident, and that the pattern of its victims suggested that it passed from person to person.

King Charles II was impressed by this work, especially the new understanding it brought of the much-feared plague. John Graunt became a Fellow of the Royal Society, a new and already prestigious organisation of scientists. But, although he laid the foundations of modern statistics, taking raw records and turning them into a resource that answers your questions, his glory was short-lived.

Plague returned to London in 1665, followed in 1666 by the Great Fire of London, which destroyed Graunt's haberdashery business. His lowly background and Catholic beliefs meant he'd never been fully accepted in the Royal Society, and he was pushed out. For a long time after his death in poverty, even his book was attributed to another man, William Petty.

---

[*] Or the mean, to be exact. Sometimes 'average' is used for the median, or middle value. For example, 'average earnings' usually means the earnings of the middle person, if you imagined lining up all the people in the country in order of earnings, and counting to the exact halfway person.

Which would be quite impractical, even assuming we were all honest about how much we earn. But I think you get the idea.

Median = middle. Mean = added up and split equally, like a restaurant bill when there are no penniless students saying, 'but I didn't have a starter'.

Heads, tails and happiness

The next man to take an interest in the sex ratio of infants was a Scot, John Arbuthnot, also a Catholic and a Fellow of the Royal Society. Arbuthnot worked with Jonathan Swift to create the satirical character John Bull, the archetypal Englishman, and his brother was killed fighting for the Jacobites on the side of Catholic King James II, but he was still given the job of physician to Queen Anne in 1709.

The Royal Society published Arbuthnot's paper, *An Argument for Divine Providence, taken from the constant regularity observ'd in the births of both sexes* in 1710. Using the London birth records for the preceding 82 years, Arbuthnot noted that male births exceeded female ones in each of those years.

Having previously published *Of The Laws of Chance*, his translation of a work by Huygens and the first English-language text on probability, he naturally thought in terms of gambling. Arbuthnot imagined that the sex of each baby has a 50:50 chance of being male or female. In any given year, there may be fewer boys or more boys.[*]

Suppose you assume a coin is fair, but it comes down tails five times in six. Are you just unlucky, or have you been cheated with an unfair coin? If you can calculate how likely a fair coin is to give you such a disappointing result[†], you can also work out how likely your disappointment is to have been caused by an unfair coin.

---

[*] Or, in theory, exactly equal numbers, but that's very unlikely.

[†] Throwing a coin six times gives 64 different possible results ($2 \times 2 \times 2 \times 2 \times 2 \times 2$ or 2^6), if we care about the order of heads and tails. Of those, only six fulfil our condition of one head and five tails. So the probability of getting such an unlucky result is 6 in 64, or 3 in 32 if you prefer. Which is unlucky, but not so very unlikely.

If you spent an entire evening tossing the same coin in groups of six throws, you'd expect to get that result nearly one time in 10. And for your friends to say you need to get out more.

Taking the probability of a 'more boys' scenario as half,[*] he calculated the probability of 82 successive years of more boys, if the underlying ratio is 50:50. It's the same probability as tossing a coin 82 times and throwing 82 heads in a row.

The chance of getting heads 82 times out of 82 is one in 4,835,703,278,458,516,698,824,704. Very unlikely indeed. Probably not a fair coin.

So Arbuthnot rejects the hypothesis that the difference he observed is down to pure chance, and asserts that it must be divine providence at work.[†] This approach, to test a null hypothesis of no underlying difference, by calculating how likely you would be to see your results if the null hypothesis were true, is widely used in science today.

Another Scotsman introduced the word statistics to the English language. Sir John Sinclair, 1st Baronet of Ulbster, wrote his *Statistical Account of Scotland* in 21 volumes between 1790 and 1799. Sir John heard the word statistics while travelling in Germany,

> *Though I apply a different meaning to that word – for by 'statistical' is meant in Germany an inquiry for the purposes of ascertaining the political strength of a country or questions respecting matters of state – whereas the idea I annex to the term is an inquiry into the state of a country, for the purpose of ascertaining the quantum of happiness enjoyed by its inhabitants, and the means of its future improvement.*

Quantum Of Happiness. Great name for a spy thriller – I should write that next. Anyway, Sinclair's desire to use his information

---

[*] Just under, including the chance of exactly equal numbers.

[†] By the time the babies reach an age to marry, the numbers have evened up, which Arbuthnot took as evidence that God weights the dice to compensate for more boys dying early.

for the future happiness of the Scots is evident in the 160-question form he asked all 938 parish clergymen to complete:

> 159. Do the people, on the whole, enjoy, in a reasonable degree, the comforts and advantages of society? And are they contented with their situation and circumstances?
>
> 160. Are there any means by which their condition could be ameliorated?

These busy men were not prompt to complete Sinclair's extensive questionnaire, covering the population's common illnesses, religion, occupation and age at death, but also geography, climate, agricultural production, provision for the poor and the price of fish. It took nine years to gather all the information, so it's not a census in the modern sense of the word.

Sinclair went on to become the oldest founder member of the Statistical Society of London (now the Royal Statistical Society) in 1834.

For a more successful early attempt to gather data on a population, let's take a look at Sweden, where astronomer Pehr Elvius also took an interest in the number of babies being born, though for different reasons.

Elvius – the early numbers

Pehr Elvius was appointed to the Swedish Royal Academy of Science in 1744, and in the same year he published an article in their journal under the title, *Catalogue of the annual number of children that are born in U—— town during the last 50 years, with reasons for remarks upon it.*[*] The coy initial disguising Uppsala town was to baffle Sweden's foreign enemies, from whom Elvius was keen to disguise the size and health of the population.

---

* Still not that snappy.

Sweden had good records of births and deaths, kept through the church, which was closely linked to the state. In fact, the involvement of the Swedish clergy in gathering information had a long and complicated history. Until the late seventeenth century, the parish priest was also responsible for registering parishioners for taxes and military service, a role that could have been specially designed to make him unpopular.

By the early eighteenth century, although each diocese had records of births and deaths from all its parishes, these records were not always kept in the same format, or compiled centrally. But in the 1730s there was rising interest in using the data. Bishop Erik Benzelius presented some of the figures to parliament in 1734, a national health board with responsibility for development of the population was set up in 1737, and the Royal Academy of Sciences[*] was founded in 1739.

Sweden was, at the time, a largely agrarian country, and had recently lost a war, and with it some provinces in northern Europe. So they had reason to worry about the size and vigour of their population.

The first Swedish census, or Tabellverket, was taken in 1749 by parish priests, completing the forms designed by Pehr Elvius and three other members of the Royal Academy of Sciences. However, reality proved messier than the forms. Parish priests were used to recording births, marriages and deaths, but less happy with age and occupation.

The annual census was reduced to being a three-yearly and then a five-yearly event. Pehr Wargentin, another astronomer who succeeded Elvius as Secretary of the Tabular Commission,

---

[*] Yes, the same one that gives out Nobel Prizes for science today. In spite of its name, the Royal Academy of Sciences was an independent body, set up by scientists, merchants, civil servants and politician Count Anders Johan von Höpken, a founder of the Hat Party.

noted that teenagers were implausibly likely to be recorded as under 15, and older citizens to be over 60. Only those aged 15–60 were liable to pay tax.

The urge to count the people of Sweden, to monitor their marital habits and to predict their future childbearing, survival and productive work came from the desire to shape the country's future wealth. Borrowing the term 'political arithmetic' from Englishman William Petty,* researchers quantified everything from people to pigs in specifics that William I of England would have envied. Not only did they calculate that a woman could count as three-quarters of a man, in terms of work, but they also predicted how large a population each parish could support.

This period of Swedish history is known as the Age of Freedom, but by today's standards that freedom was limited for most of the population. The constitutional monarchy, a parliament dominated by aristocrats and wealthy merchants, and the strong social and moral power of the church ran the lives of the masses. The vision towards which they were herding the people was one of prosperity, peace and an expanding population. Wargentin even suggested that emigration should be made a criminal offence.

Sweden led the way in statistics, in a state counting and measuring its own population. But many of the underlying ideas came from wider Europe, where the sense was growing that, by applying science and reason, society and people themselves could be made better. And it is to Enlightenment France that we go next.

Laplace's Demon

The different numbers of boy and girl babies also attracted the interest of Pierre Simon Marquis de Laplace, a Frenchman who was smart enough to combine astronomy,

---

* The same one who got the credit for Graunt's book.

physics, mathematics and staying alive during the French Revolution.*

As the title of his first book, *Exposition du Système du Monde* (an exposition of the system of the world) suggests, Laplace was confident that science and mathematics could explain everything. And by everything he meant not only planets, light and heat, but also human population and even the likelihood that a jury will reach the correct verdict.

Laplace believed firmly that:

> *We may regard the present state of the universe as the effect of its past and the cause of its future. An intellect which at any given moment knew all of the forces that animate nature and the mutual positions of the beings that compose it, if this intellect were vast enough to submit the data to analysis, could condense into a single formula the movement of the greatest bodies of the universe and that of the lightest atom; for such an intellect nothing could be uncertain and the future just like the past would be present before its eyes.*

This theoretical, all-knowing intellect has become known as Laplace's Demon. The deterministic view of the world it expresses, a secular version of the orderly universe controlled by an omniscient deity, gained popularity after the Enlightenment. It still convinces some people today, though it's hard to reconcile with the idea of humans having free will. If everything in the future is determined by what happened in the past, that leaves no room for us to make choices.

Laplace did think people are determined by their past, so his theoretical Demon would know what we will do in the

---

* He wasn't a marquis until long after the revolution, when he'd also survived the rule of Napoleon Bonaparte and the Restoration of King Louis XVIII. He was, however, a member of the Royal Academy of Science and a teacher at the Royal Military School, and plenty of other scientists lost their heads at this time.

future. But he was also a sensible man with enough worldly wisdom to keep working throughout three major regime changes and evade the guillotine. Laplace recognised that a mere human, even one as clever as him, couldn't achieve this kind of objective, all-knowing certainty about the future. The best he could hope for was to measure his own ignorance, and calculate the odds that different versions of the future would turn out to be true.

He wasn't particularly interested in babies – it's just that both France and England were keeping good records of births, which provided masses of raw material. What really interested Laplace was how to start with observations from nature and work back to the causes of things. Specifically, to work out what causes the movements of stars and planets, by observing the heavens. How could he use recorded measurements to understand why things happen?

When Laplace observed that more boys than girls were born he asked himself, like Arbuthnot: How likely is it that I'm just seeing natural variation at work?

Laplace used an approach first invented by English nonconformist clergyman Thomas Bayes in 1764, though he didn't find out about Bayes's work for some years.[*] Starting with the same question as Arbuthnot, Laplace asked: If I assume that, in the long term, exactly equal numbers of boys and girls are born, what are the odds that I would get the results these birth records are showing me?

Like Bayes, Laplace read *The Doctrine of Chances*, a contemporary book on gambling by Abraham de Moivre. Both Bayes and Laplace made the same leap of imagination.

You can put a number on how likely you think you are to win in the future: one in six for a fair dice, for example, or a 50–50 chance a coin will come down heads. Can you also put

---

[*] When Laplace did find out, he gave Thomas Bayes credit for the discovery, and this kind of approach is still known as Bayesian. Expressed in mathematical form, it's called Bayes Theorem. But Laplace did much more work to turn it into a usable method.

a number on what happened in the past, based on what you know about the results? Not a definite number, perhaps, but a range of likely numbers, with some idea of just how likely that range is to contain the true answer.

It's a tricky idea, and Laplace wrestled for a long time with turning his hunch into useful rules and methods. His fundamental approach, starting with a prior probability – 50:50, for example – and using the data you observe to change that to the most likely posterior probability, is still in use today, bearing Bayes's name.

Using these methods, Laplace showed not only that the underlying birth ratio in Paris was extremely unlikely to be anything except 'more boys', but also that the ratio in London was very probably even more male-weighted than in Paris.

But what Laplace really wanted was a numerical description of how far wrong his results were likely to be, bringing together the idea of probability, or chance, with observations of the real world.

His *Analytical Theory of Probability*, published in 1812, developed the idea that the observed birth rates, for example, will vary from the underlying average according to a predictable pattern. Though the difference between real life and his theoretical model would look random to the human eye, even if the deterministic demon knew how every last atom of the universe must fall into position.

Randomness, against everything we naturally expect, follows predictable patterns. Surprisingly, for large enough numbers of observations the emerging data will follow a similar pattern to what you'd get by tossing a coin.

If you repeatedly toss it 10 times, the expected value, five heads, will be the commonest result. Less likely results appear less often. Even if you had no idea beforehand that a coin is equally likely to come down heads or tails, you would eventually infer that it must be so, by looking at the overall spread of results.

And the more often you repeat an experiment, or observe the world, the more dramatically the results converge on the true underlying average value.

Laplace's star pupil, Siméon-Denis Poisson, developed this work further, publishing *Poisson's Law Of Large Numbers* in 1835. Their assertion that human events such as suicides can be studied in the same way as the movements of planets was controversial at the time, but their results did correspond to what was recorded in official statistics.

Average man – are you normal?

While Laplace and Poisson* were working in Paris, they met a young Belgian, visiting the city to study astronomy.

Adolphe Quetelet had shown early interest in painting and sculpture before becoming a mathematics teacher in Brussels. He had published poetry and translated Romantic writers Byron and Schiller into French, but meeting the French mathematicians set him on a new course in understanding mankind.

Quetelet continued his career in astronomy, but his use of statistics to study humanity went much further than Laplace or Poisson. He collected figures on everything from height and weight to age at marriage or penchant for committing crime.

If you've ever calculated your Body Mass Index (BMI), you're using Quetelet's work. Known until 1972 as the

---

* Poisson also gave his name to the Poisson Distribution, which describes how many rare events occur in a particular period of time, if we know the overall rate at which they happen, though each individual event is random.

For example, statistician Ladislaus Bortkiewicz counted the number of Prussian cavalrymen killed by the kick of a horse over 20 years. In each army corps the number of such deaths varied between none and four each year. In any given year, over half the corps had no deaths at all.

Independently of Poisson, Bortkiewicz spotted the same pattern of variation, and published a paper on it, confusingly called *The Law Of Small Numbers*. Some people claim that we should be talking about the Bortkiewicz Distribution, not the Poisson Distribution. I predict that if we did so, the variations in spelling would take on a much wider and more random distribution.

Quetelet Index, it is taken directly from his observation of the average relationship between height and weight.

Quetelet studied variations between individuals and across time, finding that both physical and social characteristics often fall into the same pattern around a characteristic value, if the numbers are large enough. This shape, known as the normal distribution or bell curve, is the same one used by Laplace and Poisson to describe how measurements vary around an underlying average.[*]

In 1831, Quetelet published two pamphlets, one on variations in bodily measurements, and the other on crime. In the second, he spelled out the implications of studying populations instead of individuals:

> *The greater the number of individuals, the more the influence of the individual will is effaced, being replaced by the series of general facts that depend on the general causes according to which society exists and maintains itself.*

What would Quetelet's erstwhile hero, 'mad, bad and dangerous to know' poet Byron, author of *Don Juan*, have made of this usurping of individual will by general facts and general causes?

Quetelet's major work, published in English as *A Treatise on Man* in 1842, introduced the term 'social physics'. It was a provocative parallel between the mass of humanity and the deterministic behaviour of heavenly bodies and forces that he studied in his Brussels observatory. He also coined the term '*L'homme moyen*': average man.

The idea that crime, for example, is a product not of individual moral failings but of environmental conditions and influences, was central to his work. 'Society prepares the

---

[*] It's also sometimes called the Gaussian distribution, after mathematician Gauss. Not to be confused with the Poisson Distribution, which looks less like a bell and more like a ski slope.

crime, and the guilty person is only the instrument by which it is executed.'

One consequence of this is the idea that, by changing the circumstances in which people live, you can reduce the chance they will commit crime. Social reformers have worked on this basis for centuries. British Prime Minister Tony Blair promised to be 'tough on crime, tough on the causes of crime', seeking to balance the pledge of justice – punishing those who have broken the law – with the idea that crime has causes beyond the individual.

Quetelet himself tried to distinguish between different types of cause: an individual's 'tendency to marry', for example, and the opportunity he has, or doesn't have, to fulfil that desire. He was careful to say that a person's predispositions, their environment and chance all have an influence on their actions.

For Quetelet, the Average Man was no figure of speech, but some kind of ideal that Nature aimed to produce with every individual. Deviations from this ideal were down to the influence of chance elements, which explained the fact that real people vary from the average in a similar pattern to such truly chance events as tossing the same 10 coins over and over again.

Whatever the cause, it remains true that many human characteristics do vary in this predictable pattern around an average value. Deviations from the normal distribution may genuinely be a sign that something interesting is going on.

Quetelet himself looked at the height records of 100,000 conscripts to the French army in 1817. As expected, he found that their heights varied, with over a quarter falling between 1.597–1.651m (5ft 3in–5ft 5in), and[*]

---

[*] Records suggest that Frenchmen at the time were around 7.5cm (3in) shorter than their English counterparts. This may have been due to worse nutrition or general health. Intriguingly, studies of Englishmen recruited to the East India Company found that literate recruits were around 6.35mm ($^{1/4}$in) taller than illiterate recruits. So perhaps reading this book will make you grow taller?

fewer in the other height categories, with numbers of men decreasing as the heights get further away from the mean. Except, that is, for the lowest category, those shorter than 1.57m (5ft 2in). There were 28,620 of these short men, far more than Quetelet expected from his mathematical model.

However, he didn't spend long wondering about the cause: men in this category were too short to be eligible for military service. And sure enough, there were fewer men than he expected in the two height categories above the cut-off measurement. In short, the French army had lost a couple of thousand men to strategic slumping, hunching and bending of the knees.

Florence Nightingale was very influenced by Quetelet and his book, corresponding with him and calling him the founder of 'the most important science in the world', statistics. As tutor to Albert of Saxe–Coburg, Quetelet had an enduring influence on the man who later became Prince Albert, the consort of the British Queen Victoria. Albert introduced Quetelet to the pioneer of mechanical calculation, Charles Babbage, whom we'll meet in the next chapter, and to the Reverend Thomas Malthus.

Common census

In 1799, the Reverend Malthus travelled through Sweden and observed that people were padding out their bread with tree-bark. In a revised 1803 edition of his *Essay on the Principle of Population*, he linked this starvation to the increase in Sweden's population from 2,229,661 in 1751 to 3,043,731 in 1799.

He took his figures from the comprehensive records in the Tabellverket, but drew the opposite conclusion from the Swedish statisticians: that a growing population is a bad thing, and leads to less wealth, not more.

Malthus sought to show that an increasing population must inevitably lead to disease and famine, a pessimistic view of the human future that persists to this day, in spite of a world population that has vastly outstripped every doom-laden prediction from Malthus onwards.

The first proposed census in Britain was rejected in 1753, partly on grounds of superstition. In the Bible, King David orders a census, which is followed by a plague.

But there were also political reasons. The population was unwilling to be scrutinised, fearing it would lead to more taxes and military conscription, with one member of parliament declaring the project

To be totally subversive of the last remains of human liberty …
The addition of a very few words will make it the most effectual
engine of rapacity and oppression that was ever used against an
injured people … Moreover, an annual register of our people will
acquaint our enemies abroad with our weakness.

It may have been the last point that clinched the argument, and the project was dropped.

But when Malthus's essay was first published in 1798, it fed growing fears in Britain that the population was outstripping its capacity to produce food. A bad harvest in 1800 stoked fear of starvation, or perhaps of hungry rioting masses.

The first UK census was authorised in the 1800 Population Act, and taken in March 1801. Parish officers and other worthies counted the number of houses, men and women, and their general area of employment: agriculture or manufacturing. The clergy reported the number of baptisms, marriages and deaths.

Ten years later, the exercise was repeated, with a few extra instructions, such as recording people's ages in five-year groups, and using blotting paper when taking records in ink. With one exception, Britain's population has had a census every 10 years since 1801.

People weren't always happy about being thus surveyed.

In 1911, women campaigning to be given the vote organised a mass boycott, declaring, 'women do not count, neither shall they be counted'. Some of them stayed away from home on the night of the census, listening to an Ibsen play in Portsmouth or staying in horse-drawn caravans on Wimbledon Common. By hiding in a Westminster broom cupboard, Emily Wilding Davison recorded her address as 'the Houses of Parliament', an act now commemorated with a plaque inside that cupboard.

As in Sweden, the parish registers kept by the priests were not entirely satisfactory. Since 1753, marriages were only legally binding if conducted within the Church of England. Many Nonconformists, such as Baptists, got married in their local Anglican Church, but Roman Catholics sometimes chose to marry illegally within their own church and thus evaded the register. Baptisms or circumcisions, and burials, were also conducted in other congregations. The Parish Register Act of 1812 tried to tighten things up, but a growing, industrialising nation needed more.

Thomas Henry Lister's chief contribution to the world of literature is the 1826 romantic novel *Granby*, in which the eponymous hero's love for Miss Jermyn overcomes her parents' opposition when he is revealed as the heir to Lord Malton. Which seems scant qualification for his being the first Registrar-General of Births, Marriages and Deaths of England and Wales. Nevertheless, Lister took up this post in 1836, and set up the General Register Office.[*]

From 1 July 1837, all births were to be registered, though some parents failed to do so, either from negligence, or to avoid compulsory smallpox vaccinations. Deaths were to be reported by next of kin, or whoever was present at the death or found the body, and marriages by the officiating minister of whatever religion.

Next, Lister turned his attention to the census, next due in 1841. For the first time, every household received a form, and

---

[*] Scotland, though included in the UK census, did not get civil registration of births, marriages and deaths until 1855, when William Pitt-Dundas was appointed as the first Registrar-General for Scotland. He is not recorded as having written any romantic fiction whatsoever.

was asked to report not only the number of people in the house on a specified date but also their names, occupations, and birth parish. Soldiers and sailors were now included, along with 5,016 persons travelling on trains at the time of the census.

This was also the last time that the parish registers were included in the report: the General Register Office was now a more reliable and comprehensive source of information.[*]

Death by worms

The job of turning the census returns into a useful document fell to William Farr, who classified occupations and related them to the death records, an enormous task at a time when paper records had to be compared and transcribed by hand. He noted that miners 'die in undue proportions', and that tailors were dying in surprising numbers between the ages of 25 and 45.

Farr was a qualified doctor with a keen interest in public health, and saw the importance of a standardised classification of causes of death based on logical principles, not the prevailing alphabetic system running from 'abortives' to 'worms'. However, it took years of argument before the International Statistical Institute agreed the first *International List of Causes of Death* in 1893.

The list in use today is the tenth revision, and includes entries such as 'X24 – contact with centipedes and venomous millipedes (tropical)' and 'N48.3 – Priapism'.

Farr was not only concerned with death. His annual reports as Compiler of Abstracts, later Superintendent of Statistics, at the General Register Office also covered topics such as the different factors that affect one's likelihood of getting married,

---

[*] Meanwhile, in the US, the first state law requiring registration of deaths was passed in Massachusetts in 1842. In spite of the American Medical Association urging other authorities to follow suit, national coverage wasn't attained until 1933.

such as war, abundance and wage levels. He developed life tables to predict the life expectancy of different groups, and designed the Post Office insurance scheme. He worked with Florence Nightingale on army hospital reform, on the Indian Sanitary Commission and the Royal Commission on Mines.

However, Farr did spend a lot of time analysing the causes of death. Not only the direct, proximate reason why an individual has departed the world, but also the factors that might have hastened that departure, and could possibly be prevented, or at least reduced.

One focus of his keen analytical eye was cholera, which killed thousands in Britain during the nineteenth century after arriving from India in 1831. After a second outbreak ravaged London in 1848–49, Farr collected data on where the deaths happened, and on factors that might explain the difference in death rates between districts.

South London was worst hit, with nearly eight deaths per thousand residents, compared to little over one death per thousand in north London. Farr gathered information on the density of the population in each district, measures of poverty and the underlying annual death rate. But the measure that most interested him was elevation above the high-water line of the tidal river Thames.

The prevailing theory of the time for how cholera spread was miasma: bad air, rising from the river, and carrying the illness with it. And the river Thames in the early nineteenth century was truly foul. The growing city's human and animal sewage, and other rubbish, all ended up in the Thames, untreated, where it festered until the tide took it out to sea. The river was wide, with shallow, sloping banks, so the raw sewage probably had a few days in which to achieve maximum ripeness under the noses of the unfortunate Londoners.

Farr analysed his data for the 38 registration districts of London, combining the districts according to elevation above high water, in 3m (10ft) categories. The results were striking. The mortality rate for cholera in 1849 was highest within 6m (20ft) of river level, with over 10 people per thousand of the population dying of cholera that year. Move up to a district 9–12m (30–40ft) above the stinking Thames, and the death rate falls to around six per thousand, and so on, with the improvement becoming more gradual in higher districts. Above 104m (340ft), less than one person in a thousand died of cholera.

Conclusive proof, as far as Farr was concerned, that the miasma theory was correct, and that eliminating the noxious gases would reduce the dreadful death toll. And if you saw the results as an infographic in a modern newspaper, you'd probably agree.

The relationship between elevation and your odds of dying of cholera was large, consistent, and showed a greater effect for a greater exposure to the risk factor: in this case, being low down where the foul air lay. It also fitted current scientific theory.

However, another doctor had a conflicting theory. John Snow, a physician living and working in Soho, published *On The Mode of Communication of Cholera* in 1849. Snow had first encountered the disease in Newcastle when, apprenticed to a surgeon, he treated patients there in the 1831–32 epidemic.

His accounts of how whole families were wiped out within days are terrifying, and call to mind the Ebola outbreak ripping through parts of Africa as I began writing this book, with the same factors of poverty, people living in close quarters with inadequate water supplies, and failing to dispose of contaminated belongings.

Closely observing how the disease appeared to spread from patient to patient within families, or via clothes and bedding belonging to somebody who had died of cholera, or to people who washed and laid out a body, Snow developed the theory that it was spread via the alimentary canal. To put it bluntly, the diarrhoea that killed one cholera patient got into the mouth of the next victim, one way or another.

Patterns in the way the cholera epidemic spread suggested to John Snow that just washing the bedlinen used by one patient could be enough to pollute a water source, and that everyone who subsequently drank from that source was at risk of catching the disease. Observations in overcrowded courts around London, where waste water leaked into the well or spring from which people drank, reinforced his hypothesis that water transmitted cholera.

In 1854, Snow was living at 54 Frith Street, Soho Square. On 3 September, he heard of a sudden upsurge in cholera cases in nearby Broad Street – 83 deaths registered within three days. His suspicion fell on the water pump from which many local homes, pubs and coffee shops got their drinking water, but when he looked at the water, it appeared clean.

Nevertheless, seeing no other likely cause for such a violent outbreak he continued his investigations. Taking copies of the death registrations for the period, he found that most of the victims lived in houses for which the suspected pump was the nearest water supply. Of the 10 victims who had an alternative source of water closer at hand, eight were known to drink from the Broad Street pump.

The map on which Snow marked deaths by cholera with black bars has become one of the classic documents of epidemiology. The darkening of the map around the guilty pump is striking. However, the relationship between closeness to the pump and odds of dying is not entirely simple. For example, the brewery near the pump in Broad Street employed over 70 workmen, and not one of them died of cholera. The workhouse in Poland Street, whose crowded conditions and underfed inmates were surrounded on all sides by infected houses, lost only five inmates out of 535. If the workhouse had reflected the death rates of the surrounding streets, over 100 of them would have been dead.

Snow's careful work revealed that both brewery and workhouse had their own well, strengthening his case that water supply, not mere proximity, was the main risk factor. The brewery proprietor, Mr Huggins, told Snow that his men got an allowance of malt liquor, and he did not believe they drank water at all.

To add more strength to his argument, Snow found specific cases where individuals who passed through the area just long enough to eat and drink subsequently died of cholera. He even found a Hampstead lady who died after drinking Broad Street pump water, which she had delivered daily because she preferred the taste. Her niece, visiting from Islington, went home and died from cholera. Neither Hampstead nor Islington had any other cases at the time.

On 7 September, Snow presented his evidence to the local authority of its day, the Board of Guardians of the Parish of St James, and requested that the pump handle be removed to prevent any further deaths caused by drinking the contaminated water. The guardians, while not entirely convinced by Snow's theory, must have thought it was worth a try, because they agreed.

Since so many had already died, or fled the area, the outbreak may already have been coming to an end, but the removal of the pump handle is commemorated every year with a Pumphandle Lecture, followed by a drink in the John Snow pub in Soho.

However, neither William Farr nor general scientific opinion was convinced by Snow's theory, in spite of the fact that Farr's data on the 1848–49 cholera epidemic in London, which had apparently confirmed the miasma theory, also included information about the sources of water in the different districts he studied.

Water supplies in London were provided by private companies, and if you weren't lucky enough to have your own well or pump, your drinking water would be piped in from the Thames or one of its tributaries. This meant that most of south London was drinking water drawn from the Thames between Battersea Bridge and Waterloo Bridge, downstream of the places where sewage was flowing into the river. Coincidentally, most of these districts were also low-lying.

So the clear link that Farr saw between low elevation alone and high risk was erroneous; low elevation did not cause the disease, or mean that one was at a greater risk of catching it.

If, instead of sorting the districts by elevation, he had sorted them by main water supply, he would have seen that the difference between those categories was stark. In districts getting their water from the Thames above Hammersmith, cholera killed between 11 and 19 per 10,000. Among those taking their water from the Thames below Battersea Bridge, between 77 and 168 people per 10,000 died from cholera.

In the long run, the story has a happy ending. The stench of the Thames became unbearable. The Houses of Parliament, lying alongside the river, were so filled with the Great Stink in the summer of 1858 that MPs were driven out of the building, and swiftly approved a bill to create a new sewer system for London. Joseph Bazalgette's ambitious scheme gathered all the city's sewage and piped it far downstream, separating drinking water from human waste.

By ending the stink, wrongly blamed for spreading disease, the Victorians inadvertently solved the real problem: contaminated drinking water.

An unhealthy legacy

'The word Eugenics,' begins a *Jewish Chronicle* article from 1910, 'will be for ever associated with the name of Sir Francis Galton, who has devoted a long life to the pursuance of a high ideal – that of improving the fitness of the human race ...'

That may have been true then, when Francis Galton was still alive, aged 89. Today, the word eugenics has more chilling associations.

Back in 1910, eugenics was a fashionable and popular idea with people who considered themselves progressive. Socialist writer George Bernard Shaw, while strongly against state-enforced eugenics, believed that social reforms would eventually lead to selective breeding of better human beings. Some campaigners for the availability of birth control were motivated not only by the desire to enhance women's reproductive freedom, but also to reduce the population among certain sectors of society.

Charles Davenport wrote to Francis Galton from America in October 1910, telling him 'the seed sown by you is still sprouting in distant countries'. Davenport had opened the Eugenics Record Office (ERO) at Cold Spring Harbor in Long Island, New York, and published *Eugenics: The Science of Human Improvement by Better Breeding* that year. Davenport's successor Harry Laughlin used data collected by the ERO to campaign for public policy including compulsory sterilisation and restrictive immigration laws, some of which were used as models by the Nazi regime in Germany.

The Eugenics Record Office closed in 1939, but several states had already passed laws based on Laughlin's recommendations. Sir Winston Churchill was strongly in favour of 'the improvement of the British breed' by segregation or sterilisation of the 'feeble-minded', and Canada established the Alberta Eugenics Board in 1928 to sterilise 'mentally deficient' individuals against their will, under an act not repealed until 1972. Australia, Iceland, Norway, Sweden and Switzerland are among other countries that have sterilised the mentally ill or handicapped.

Today, alongside its educational work in genetics, the Cold Spring Harbor Laboratory sells books, including *Murderous Science: Elimination by Scientific Selection of Jews, Gypsies and Others in Germany 1933–1945* and *The Unfit: A History of a Bad Idea*.

But when Francis Galton, a cousin of Charles Darwin, coined the term in 1883, he described it as 'the study of the agencies under social control that may improve or impair the racial qualities of future generations, either physically or mentally'. He would have seen himself very much as a scientific, rational man of his time, looking, like Quetelet, for ways of harnessing science to improve the future of humanity.

Galton was no great mathematician. He qualified as a doctor, but then an inheritance allowed him to give up medical practice and explore whatever interested him. Initially, that meant Africa. Next, he turned to the study of the weather, collecting meteorological data from across Europe for December 1861, and creating visual charts that enabled him to look at many different variables at once. By comparing wind direction, temperature and pressure, he discovered the anticyclone.

When his results were published in 1863 as 'Meteorographica', Galton included a note of warning to those about to be dazzled by what today we'd call an infographic: 'it is truly absurd to see how plastic a limited number of observations become, in the hands of men with preconceived ideas.'

With his illustrious cousin Charles, Galton shared an illustrious grandfather: Erasmus Darwin, poet and scientist. In fact, the family produced an impressive number of prominent men, which may have led Galton to wonder whether illustriousness was an inherited trait like height. Galton made this comparison explicit in his book, *Hereditary Genius*, in which he developed some of Quetelet's ideas about average man to look at individuals instead of populations.

Although an understanding of *how* genetics worked was still years away, people had begun to observe that inheritance could be mathematically predictable. Not only individual traits such as colour blindness, but qualities that are continuously variable, such as height, follow mathematical patterns in the population. And observations made about animals or even plants could also apply to human beings.

Galton was interested in how qualities such as height were passed down from parents to children. Bribing his experimental subjects with the chance to win money, he collected lots of data on the heights of parents and children, adjusted for the fact that women tend to be shorter, and averaged the parents' height to get a theoretical 'midparent' against which to compare the children.*

---

* Obviously the children were grown up. Comparing eight-year-olds with their parents wouldn't be much help.

He was looking for a clear relationship between parents' heights and children's heights, and by measuring 928 adults and their parents, he found it. But he also found something that would be much more important.

Galton noticed that, although the average heights of children followed their midparent's height quite closely, the children of very tall or short parents were not as extremely tall or short themselves.

Having read Quetelet's work in 1863, Galton was familiar with the idea of the normal distribution, and made his own illustration of the Law of Deviation from an Average, showing how the heights of a million men would be grouped, mostly within a few inches of the average, with fewer and fewer individuals at the extreme ends of the scale.

Now Galton had discovered a principle that's so deceptively simple it's surprising how often we forget about it. Things slide back towards the middle, like people sharing a hammock rolling into the centre.* Today we'd call it regression to the mean, or going back to the average.

One current example is speed cameras, which are put up on dangerous stretches of road after a run of accidents, to slow down traffic and prevent future collisions. Accident rates do tend to fall after a speed camera is installed. The problem is, you'd expect the number of accidents to go down anyway, whether or not a speed camera goes up.

Don't believe me? Let's think about earthquakes instead. According to the US Geological Survey (USGS), there are 16 earthquakes per year of magnitude 7 or above, worldwide. And with two offices in California, I assume they're paying very close attention. In 2010, however, the USGS recorded 24 of them. A worrying rise! Was Max Zorin carrying out his evil plot to destroy Silicon Valley by triggering the San Andreas Fault?

---

* We refer to extreme values, far from the mean, as outliers, so we can imagine them as sleepers balanced precariously on the edge of the hammock, or fallen out and lying on the ground.

Fear not, the USGS must have installed their top-secret earthquake-prevention device the following year, because by 2012, the number of magnitude 7+ earthquakes was down to only 14, two below average.

Of course, they don't really have an earthquake-prevention device. Max Zorin is James Bond's nemesis from 1985 film *A View to a Kill*, and his fictional plan is as gloriously silly as any other Bond villain. Nothing sinister is going on, just natural variation around an average. The 10-year high in 2010, a year I chose because it *was* a 10-year high, was followed by lower counts, just as you'd expect.

And in the same way, a four-accident high on our fictional road would usually be followed by a fall back towards or below the average value. Research suggests that speed cameras can have some effect in reducing accidents, but less than you might think by looking at the figures from just before and after their installation.

What do speed cameras and earthquakes have to do with Galton's children and mid-parents I hear you ask? If the underlying tendency is to be the same height as your midparent, but a series of chance factors all contribute to making you much taller, it's unlikely that your own children will be dealt exactly the same hand of genetic and environmental cards. If you think of very tall, or very short, parents as extreme variations from the average, then you'd expect them to be followed by a return towards more common values. Which is what Galton observed.

However, this 'regression towards the level of mediocrity' was only the first step towards what Galton was really looking for. He wanted to know whether heredity could be expressed mathematically. As he put it himself:

> *Given a man of known stature, and ignoring every other fact, what will be the probably average height of his brothers, sons, nephews, grandchildren, &c., respectively, and what proportion of them will probably range between any two heights we please to specify?*

Applying the patterns from a population to predictions about individuals is one of the oldest problems in data, or statistics. As Galton put it: 'Whatever is statistically certain in a large number is the most probable occurrence in a small one.'

If you know nothing about an individual except the population from which you've picked them, there's a simple way to calculate the probability of their height falling within a specific range, using the normal distribution, or bell-curve. If you know the mean value, and a couple of other things about how the other values spread out from it, you can calculate the chance that one value will fall within a given range.

What Galton had developed was a way to use both this data about the underlying population and the specific height of the midparent, or uncle, or brother to predict the height of the child, or nephew, or brother. Or rather, to predict how likely that height is to fall within a certain range. He called this the 'ratio of regression', based on:

> *a compromise between two conflicting probabilities: the one that the unknown brother should differ little from the known man, the other that he should differ little from the mean of his race. The result can be mathematically shown to be a ratio of regression that is constant for all statures.*

Galton calculated the probability that a son would be at least the same height as his father at 50 per cent for a father 1.73m (5ft 8in) tall, but at less than 1 per cent for a father 1.96m (6ft 5in) tall. Francis Galton called this measurable link between two sets of data 'correlation'.

Height wasn't his main interest, however. Having done all this work, he turned his interest swiftly back to studying how genius could be passed on through families. In 1874, he had conducted a survey of 150 prominent British scientists, from which he concluded that key factors were energy, general good health, independence of mind and an interest in science, but only when combined with discipline and focus.

He then called for a wider application of the statistical principles he was applying to physical characteristics in a population to character traits. 'The habit should therefore be encouraged in biographies, of ranking a man among his contemporaries, in respect of every quality that is discussed, and to give ample data in justification of the rank assigned to him.'

He even entertained himself, during a tedious talk being given by somebody else, by measuring the frequency, amplitude of physical movement and duration of fidgeting among the audience. It was an average one fidget per person per minute, in case you want to carry out a similar study next time you're at a boring meeting.

Not everybody was readily convinced that the ideas of correlation and regression could be applied to social questions. 'Personally, I ought to say that there is, in my opinion, considerable danger in applying the methods of exact science to problems in descriptive science, whether they be problems of heredity or of political economy.' These sceptical words were spoken by Karl Pearson, in his lecture on Galton's 1889 book, *Natural Inheritance*, at the Men and Women Club.[*]

Pearson's interests stretched to law, philosophy, the history of science, German language and literature. Then a professor of Applied Mathematics at University College London (UCL), in his spare time he gave talks on Marx and Martin Luther. So he was already seeking the reasons behind political and economic events. But was it reasonable to expect sociological theories to be as exact and logical as mathematical ones?

Pearson was impressed by Galton's work on correlation and regression. His initial ambivalence about transferring the methods of physics or astronomy to human beings was evidently resolved, as he became the first Galton Professor of

---

[*] Not a singles club, but a progressive political gathering that Pearson helped found, and which discussed social issues such as the roles of the sexes. Though he did also meet his wife there.

Eugenics at UCL in 1911, after leading Galton's Eugenics Record Office at UCL since 1906. Combining the urge to improve human life with excitement about the emerging science of evolution, Pearson developed Galton's methods to study more complex situations, in which numerous different causes might be at work.[*]

Speaking at a dinner in his honour in 1934, Pearson speaks of the 'culmination' of eugenics lying:

> In the future, perhaps with Reichskanzler Hitler and his proposals to regenerate the German people. In Germany a vast experiment is in hand, and some of you may live to see its results. If it fails it will not be for want of enthusiasm, but rather because the Germans are only just starting the study of mathematical statistics in the modern sense!

Words worth remembering whenever anybody suggests that some social problem could be solved, if only everybody understood mathematics better. Pearson himself died in 1936, so thankfully he never had to see the horrific outcome of the German 'experiment'.

Today, any suggestion that human beings should be selectively bred like farm animals is generally regarded as eugenics, a word with chilling echoes of the holocaust.

Even parents who want to use medical techniques to avoid passing on hereditary diseases to their children have to overcome this fear that creating 'designer babies' is the first step towards wiping out the rest. It's important to remember, I think, that there's a world of difference between loving parents wishing good health for their future child and others deciding they know what's best for your children, or for the human race as a whole, and imposing it upon you.

---

[*] He gives his name to the Pearson correlation coefficient, a measure of how closely two variables, such as height and weight, are related.

I liked their early stuff

But don't let's end this chapter on a bleak note. Statistics, which I like to regard as big data's early, acoustic stuff before they were famous, has given us many gifts. Better medicine, better food crops, even better Guinness, all owe a lot to statisticians. They could probably tell us how much, within a 95 per cent confidence interval.*

Applying mathematics to understanding the real world is an art as well as a science, and finer minds than mine continue to grapple with statistics. But today, whether or not they accept the label big data, they tend to use computers to do the heavy lifting. And that's where our little historical tour goes next: the Industrial Revolution of statistics.

---

* Which is the way statisticians describe a range of values that probably include the true answer, and how confident they are that the true value is within that range. The bigger the percentage, the more confident they are, as you'd expect.

CHAPTER THREE

Thinking machines

Britain in the nineteenth century was industrialising so fast that skilled engineers could demand high salaries. Charles Babbage complained that 'railroad mania' forced him to offer his chief assistant a big pay rise to prevent him leaving. His new offer of a guinea a day was over four times the usual craftsman's wage of around five shillings.

Why did Babbage, a Cambridge Professor of Mathematics, need to employ an engineer? Because he had a vision of mathematics, like railroads, mines and factories, as an industrial process in which machines would do the hard work.

Machines for doing arithmetic had existed for some time, though the lack of precision engineering meant they were slow and inaccurate. Babbage began work in 1821 on his Difference Engine, an elaborate mechanism calculating the most basic functions, based on adding numbers, by turning cogs. It would then give the answer through a printer. His design involved 25,000 parts. After 20 years, with only a portion of the engine built, the government withdrew their funding.

Babbage had already moved on to a more ambitious design. The Difference Engine was limited. It could only do the one type of calculation that was built into its design. Completed, it would have been a four-ton adding machine that could do less than the cheapest calculator any child takes to school today.

His new project, the Analytical Engine, occurred to him around the same time that he met Augusta Ada Byron,*

---

* Daughter of the poet Byron and of mathematically inclined Annabella Millbanke, whom Byron called his 'princess of parallelograms'. So perhaps he would have appreciated Quetelet's mathematics.

then 17, and showed her the working section of his Difference Engine. A keen mathematician like her mother, Ada was immediately intrigued by the potential and, after a short break to marry the future Earl of Lovelace and have some children, Lady Ada Lovelace got to work on inventing computer programming.

Unlike the Difference Engine, the Analytical Engine could perform different tasks, using an approach borrowed from another industrial process, the Jacquard Loom.

Joseph-Marie Charles, or Jacquard, came from a French weaving family. After his family firm went bankrupt, he designed a loom that could weave elaborate patterns repeatedly with minimal human input. Weaving was already mechanised in 1801, when he constructed his first Jacquard Loom. Now the patterns could be mechanised too. In spite of opposition from weavers who feared for their jobs, within 11 years there were 11,000 of his looms in France alone, and the technology was spreading to the British textile industry's power-looms.

Jacquard used punched cards to record the desired pattern and tell the loom what to do. Each position on the card could have a hole, or no hole. Jacquard used the binary system, for the same reason that modern computers do: on/off is the simplest unit of information. Strung together, the cards stored instructions for the image to be woven into the fabric, and were readable* by the loom without any human translation. One loom could weave any pattern in any colour, by having the cards and the thread changed.

Babbage saw that this flexibility could turn his clumsy calculating machine into something much more versatile. By using punched cards to store two sets of information, he

---

* Put simply, a hole meant that a rod could pass through it and press down the warp thread, allowing the weft thread to pass above it and be visible from the front of the fabric. No hole meant no rod, the warp thread stayed on top, and the weft thread remained hidden.

could tell the Analytical Engine not only what numbers to put in – the *variables* – but also what to do with them – the *operation*.

Making a direct analogy with Jacquard's invention, he referred to the Analytical Engine as having a *mill* and a *store*. The mill does the work of processing, according to the instructions and the initial inputs taken from the store. The results are put back into the store, in the same punched-card format.

Sadly, in spite of paying his assistant the generously increased salary, Babbage never achieved a working Analytical Engine. He built one part of it, which is on display at London's Science Museum, but nobody has yet constructed a complete working model.

Lady Lovelace's objection

This makes Ada Lovelace's work all the more remarkable. Using only Babbage's designs, she worked out in some detail what kind of tasks the Engine would be able to perform, and how to put the information into punched-card form. She foresaw that calculations too difficult for a human brain to do without mistakes could be performed by machine instead:

> *We might even invent laws for series or formulae in an arbitrary manner, and set the engine to work upon them, and thus deduce numerical results which we might not otherwise have thought of obtaining; but this would hardly perhaps in any instance be productive of any great practical utility, or calculated to rank higher than as a philosophical amusement.*

She even mused on whether such mechanical computation could be used for other types of information, just as Jacquard translated pictures into a series of holes in card. But she added

a philosophical note that sounds surprisingly modern for a
scientific paper published in 1842:

> *It is desirable to guard against the possibility of
> exaggerated ideas that might arise as to the powers of the
> Analytical Engine. In considering any new subject, there is
> frequently a tendency, first, to overrate what we find to be already
> interesting or remarkable; and, secondly, by a sort of natural
> reaction, to undervalue the true state of the case, when we do
> discover that our notions have surpassed those that were really
> tenable.*

Don't get carried away by hype, and then be too disillusioned
to appreciate its true potential, in other words.

> *The Analytical Engine has no pretensions whatever to
> originate anything. It can do whatever we know how to order it
> to perform. It can follow analysis; but it has no power
> of anticipating any analytical relations or truths. Its province
> is to assist us in making available what we are already acquainted
> with.*

But though it cannot create anything new, Lovelace did
foresee that it would force us to think afresh. Expressing our
questions in a mathematical form that an engine could
process would, in itself, lead us to new ways of thinking
about science.

> *There are in all extensions of human power, or additions to human
> knowledge, various collateral influences, besides the main and
> primary object attained.*

Lovelace and Babbage were far ahead of their time, but the
idea of using machines to store and process information fitted
the age of the telegraph, the factory and the steam engine.

Which takes us back to the census, this time across the
Atlantic.

Marking cards

In America, a census was carried out every 10 years from 1790, soon after the instatement of President Washington. By now, you won't be surprised to hear that the initial purpose was to work out the fairest way of sharing out the burden of the War of Independence. Both representation and taxation would be allocated according to population. And yes, they did also make a note of all the males over 16 who might be called on to fight.

Before the 1800 census, the American Philosophical Society, whose president was Thomas Jefferson, requested that it should include a range of other interesting questions about age, profession and country of origin. The aspiration to understand 'the causes which influence life and health' and to predict future population growth was strong among America's leading thinkers.

Their pleas were ignored. Not until 1820 did the census gather any more than the basic information needed to tax the population and draft them into the army. It's one thing for scientists to think it would be interesting to have some information on which to base their research, but it's something else to collect all that data and record it in a form that's easy to study.

Even when they started to record marital status, occupation and age, there was no easy method to summarise the findings. With the population growing almost as fast as the number of questions in the ever-expanding census, the results of the 10-yearly survey were taking longer and longer to compile into anything useful, such as tables of figures. The 1880 census took so long to tabulate that by the time it was finished it was nearly time to start working on the 1890 census. Perhaps foreseeing the day when it would take more than 10 years to hand-count the results, the Census Office advertised for an inventor who could solve the problem.

This was an opportunity for Herman Hollerith, who had helped out as a statistician on the 1880 census, and seen the limitations of relying on human beings to transfer all the information into tables.

His mentor John Shaw Billings, a doctor, suggested a card system inspired by libraries. Library cards were first written on the backs of playing cards by the post-revolutionary French librarians.* Harvard University Library adopted a card index in 1861 and Melvil Dewey[†] introduced the standardised system with his Library Bureau company, founded in 1876.

Combining the library index with Jacquard's punched loom cards, Billing suggested a card for each individual, with holes corresponding to different categories of data, such as age, race and occupation. A hole in the right position recorded that you were female or male, and so on.

Hollerith designed a machine that could both record the census responses and cross-tabulate the data. The Hollerith Desk combined a card for each record with a system of holes as used for Jacquard's loom.

Instead of pressing on a thread, when Hollerith's rod passed through a hole it made electrical contact with a little pool of mercury, completed a circuit and tripped a switch. As well as sorting the cards, it used a gear-driven counting device to add to each total, rotating dials to display the running totals above the desk, and a way to record summaries of information by punching new cards.

The 1880 census was finally finished in 1887, and Hollerith demonstrated his machine in the same year. He got the contract, and his Electric Tabulating Machine completed the 1890 US Census in three years.

This kind of fast, efficient information-wrangling had potential for all sorts of applications. Hollerith's Tabulating Machine Company, formed in 1896, held the patent for the Hollerith Desk, and he supplied his information technology

---

* When religious property was confiscated after the French Revolution the books were used to set up a system of public libraries.
[†] Dewey's library classification system is still in use, though most libraries have now transferred the information from cards on to a digital database.

to all sorts of companies and governments, including the Canadian, Austrian and Norwegian censuses, at a price.

Perhaps too steep a price.

The US Census Bureau was formed in 1902, and escaped Hollerith's monopoly by paying their own technician, James Powers, to develop a new machine. The 1910 census didn't need Hollerith's machines, and he now had competition from the new Powers company. In 1911, he merged with the Computing Scale Company and the International Time Recording Company to form the Computing Tabulating Recording Company, CTR.

Hollerith eventually retired to raise Guernsey cattle, but CTR thrived. It supplied businesses from the chemical industry to life insurance, and expanded internationally under its new president T. J. Watson Sr. In 1924 the company changed its name to International Business Machines: IBM.

The same IBM, 20 years later, would build the Mark 1 with Harvard University, the computer that Grace Hopper debugged in Chapter 1.

Breaking codes

In Britain, too, the Second World War spurred the development of computing.

British statisticians were doing lots of important work, devising tests that would be used in agriculture and science, but the British contribution to mechanised reasoning owes more to a pure[*] mathematician, Alan Turing. Drafted into the secret codebreaking establishment at Bletchley Park, he did work only recently acknowledged for its importance in deciphering German radio messages and helping the Allies to win the war.

---

[*] His mathematics were pure, as opposed to applied. Turing himself was not pure enough by the social standards of the day, and was later arrested and convicted for his homosexuality.

The challenge for Turing and his colleagues was to make sense of scrambled messages, and thus to predict what enemy forces would do next. They had some idea of how the messages were encoded, using electrical machines like Enigma, a system of wheels that could be put into many different starting positions. They even knew Enigma was designed to shift the wheels during the encryption process, so an E in the original text would be represented by different letters in different parts of the encoded text.

The problem was there were 159 billion possible starting settings. Polish mathematicians had worked out how to decode Enigma in 1932, while the German Army was still testing the technology. But after the war started, the cipher used to encode messages changed every day, instead of every few months. There wasn't time to work through all the possibilities before the code changed again, so the Polish codebreakers shared their work with the British intelligence services.

They needed machines that could test the options – secret calculating machines known as bombes to imply that the new technology was explosive, not analytical. Their banks of wheels whirred and clicked around, like giant versions of Hollerith's desk. They were mainly operated by Wrens: members of the Women's Royal Naval Service, not specially trained small birds.

But just as important as the industrial scale alphabet-crunching of the bombes was Turing's system for narrowing down solutions that would make sense. Named Banburismus, after the nearby town of Banbury, his method used cards and a few clues from what they did know to suggest the most fruitful places for the bombes to start searching. Combining this use of probability with the brute force of the bombe's wheels, the codebreakers were able to beat the odds, using the Bayesian methods developed by Laplace.

Alan Turing's role as the Father of Computing is now celebrated, long after his death in tragic circumstances. After the war, he continued to work on his personal project of 'building a brain'. His work in what we now call artificial intelligence, AI, gave rise to what's still called the Turing Test. If a machine can interact with a human being, and fool that human into thinking the machine is human[*] then, said Turing, it can be called intelligent.

But at that time, all the work done at Bletchley Park was kept secret, the hefty computing machinery broken up, the workforce forbidden to talk about their work, and all the research classified as an Official Secret. Some of the codebreakers continued their work in the new Government Communications Headquarters, GCHQ, which is still the home of the British Intelligence Services' work intercepting and deciphering telecommunications at home and abroad. Others, including Turing, found it difficult to pursue their research because they weren't allowed to discuss what they'd been doing at Bletchley.

Calculating risk

War speeded up the development of computing on both sides of the Atlantic, but the appetite for machine processing of information was already there.

Edmund C. Berkeley, an employee of the Prudential insurance company, wrote in 1939 about the potential of computing in his industry. Punch-card machines were widely used to store and retrieve data, but Berkeley had a vision of more sophisticated machines that would use mathematics to find the most efficient way to carry out a task. Not only solving a problem but also choosing the most economical way to solve it would be automated.

During the war, Berkeley worked with IBM and Howard Aiken at Harvard, where he helped develop the successor to the Mark I – the unimaginatively titled Mark II automatic sequence controlled calculator. In December 1946, he used a

---

[*] The machine is in another room, obviously. Otherwise it would also have to be a very lifelike robot, which is a whole other challenge.

similar Bell Labs machine to solve an insurance problem, looking up tables of data and using them to calculate a changed insurance premium.

This may seem a banal problem for the most cutting edge technology of the time, but insurance companies have always faced one of the hardest challenges of applied mathematics. How do you calculate the uncertainties of the future to make sure the premiums your customers pay today will cover the payouts you need to make tomorrow? Overcharge, and your customers will go to your competitors. Undercharge, and your underwriters will have to make up the shortfall, or go bust.

In the Lloyd's Building in the City of London hangs the Lutine Bell, traditionally rung when news of an overdue ship arrives. Originally the ship's bell on French frigate *La Lutine*, it sank with its ship,* and gold and silver bullion worth £1 million, off the Dutch coast in 1799. By this time the ship had been captured by the British Navy and insured by Lloyds of London as HMS *Lutine*. The underwriters paid up in full.

London wasn't the first home of marine insurance. Shakespeare's eponymous Merchant of Venice, Antonio, risks losing more than money when news comes in that his ship is 'wrecked on the narrow seas ... a very dangerous flat, and fatal, where the carcasses of many a tall ship lie buried'. But Antonio would have had better options than staking his pound of flesh.

Italian city-states of the fourteenth century introduced insurance as we'd recognise it today, sharing the risk of catastrophic loss among the merchants. The first record of a disputed insurance payment in London is a court case brought in 1426 by a Florentine merchant called Alexander Ferrantyn who had to buy back his ship, the *Saint Anne of London*, and its cargo of Bordeaux wine after it was seized by pirates.

In the early eighteenth century, Lloyd's Coffee House became the centre of marine insurance in London. Men with

---

* The bell was salvaged in 1857.

enough private fortune to risk losing some of it could become underwriters, essentially gambling that the ship they agreed to insure would return intact, in spite of weather, war and privateers. Gradually, they began to insure against other risks. The 1906 earthquake in San Francisco, for example, hit Lloyds hard.

Meanwhile, under pressure from discontented workers, American states began passing laws compelling employers to insure their employees against industrial accidents and diseases. Suddenly, the insurance industry needed to calculate how likely a Nebraskan suspender-maker* was to suffer injury or illness, and the likely cost of treatment or compensation. With limited information on previous casualty rates, they needed a better system than hunch and back-of-envelope estimation.

The Casualty Actuarial Society formed in 1914, and in 1918 it came up with a method called *credibility*. So called because it assigned a numerical value, a weight, to the reliability of data from different sources, credibility allowed actuaries to draw on all the information available at the time. It also left room to revise the calculation when new information came to light. Again, though it seems unlikely they knew it at the time, they were using the same approach as Bayes and Laplace.

Assigning the weight or credibility to each piece of information demanded human judgement. If you're establishing rates for fire protection in Oregon, then reports by fire marshals in Oregon should carry more weight than those from urban New York or industrial Pennsylvania. Introducing more advanced information technology to insurance wouldn't completely remove human beings from the picture.

---

* No sniggering on the European side of the Atlantic, now, where suspenders hold up ladies' stockings. In the US, suspenders stretch across a man's belly to keep his trousers from falling down. Or his pants, as the Americans say.

However, Mr Berkeley of the Prudential could see that using a combination of punched cards and the newer magnetic tape data storage systems could speed up some of the donkey work being done by clerical staff.

The Prudential joined the US Census Bureau to back the development of UNIVAC by one of IBM's rivals. The UNIVersal Automatic Computer, the world's first commercially available computer, sold 46 models at around $1 million each. Its internal memory held 1,000 words, but by using magnetic tape it could store as much information as required, limited only by the storage space you had available for magnetic tape reels.

UNIVAC made its television debut in 1952, with Walter Cronkite on the CBS network covering the presidential election returns. Incoming results were fed into UNIVAC, which correctly predicted that Eisenhower would win. This went against expert opinion, so CBS didn't broadcast UNIVAC's winning bet until much later, when the outcome was unequivocal.

The last running UNIVAC 1 machines, used by Life And Casualty Insurance of Tennessee, were shut down in 1970.

But UNIVAC is still a long way from the machines at work with big data today. To fill in the family tree of artificial intelligence, we need to go back to Alan Turing and his thinking machine.

Turing's child

In 1950, Turing wrote a piece for *Mind, a Quarterly Review of Psychology and Philosophy*.

'Computing Machinery and Intelligence' described the Turing Test, though he didn't call it that, by analogy. An interrogator tries to find out which of two people in the next room is a man, and which a woman, by asking a series of questions.

Turing begins by asking the reader, 'can machines think?' Attempting to answer, he describes the digital computer as being a machine that can 'mimic the actions of a human

computer very closely.' He also describes the idea of a
computer program, using the example of a mother instructing
her child:

> *Suppose Mother wants Tommy to call at the cobbler's every*
> *morning on his way to school to see if her shoes are done, she*
> *can ask him afresh every morning. Alternatively she can stick up*
> *a notice once and for all in the hall which he will see when he*
> *leaves for school and which tells him to call for the shoes, and*
> *also to destroy the notice when he comes back if he has the shoes*
> *with him.*

It's pretty obvious that Turing had no children, and spent far
too much time with very reliable people and machines.

That apart, it's a good description of how an algorithm
works. An algorithm is just an ordered set of instructions,
which can include conditional, IF, instructions. If you've ever
seen a flow chart, that's just an algorithm designed to be read
by a human being.

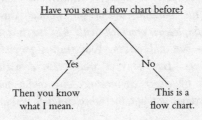

Turing also mentions Babbage's Analytical Engine, and
Laplace's view 'that from the complete state of the universe at
one moment of time ... it should be possible to predict all
future states', before making his own prediction, that by the
end of the century* it will be acceptable to talk of machines
thinking. Then he tackles various objections, including the
question of the soul, of consciousness and 'Lady Lovelace's

---

* i.e. by the year 2000.

objection' that the Analytical Engine cannot originate anything, and can only do what it is told.

It's remarkable how comprehensively Turing lays down problems that the field of Artificial Intelligence is still working on today. He's not convinced that a machine can never produce original results: he thinks a powerful enough machine could leap ahead of his limited calculations and surprise him. He agrees that it would be impossible to lay down 'rules of conduct' to tell a machine how to respond under any conditions, but suggests that instead 'laws of behaviour' could be found that govern the machine, just as 'if you pinch him he will squeak' applies to a man.

Turing even imagines machines that don't function only in absolute, yes/no terms, but can work with a range of answers, using probability to decide which are more likely to be true.

He also proposes that a machine designed to learn for itself, as a child does, could develop into something approaching an adult human brain.

> It will not be possible to apply exactly the same teaching process to the machine as to the normal child. It will not, for example, be provided with legs, so it could not be asked to go out and fill the coal scuttle. Possibly it might not have eyes. But however well these deficiencies might be overcome by clever engineering, one could not send the creature to school without the other children making excessive fun of it.

It's rather touching to think of Turing worrying about his little robot child being bullied at school for being different. Would it still remember to check at the cobbler's to see whether its mother's shoes were ready? It's also odd to find that Turing could imagine a world of thinking machines, but not one with central heating.

The term artificial intelligence, AI, wasn't coined until 1956, at a conference in New Hampshire. In 1958, Allen Newell and Herbert Simon claimed that a digital computer would be world chess champion within 10 years. Researchers were very optimistic about how easy it would be to recreate general intelligence in a machine. In 1965, a program called ELIZA carried on remote conversations, making it the first with the remotest potential to pass the Turing Test.

In the same year, Turing's wartime assistant, mathematician I. J. Good, suggested that the last invention human beings need to create is the first ultra-intelligent machine. From then on, the machines can design even better machines, and so on. This idea, of the machine that thinks better than any human, is often called the singularity today. And not everyone is so optimistic about how things would turn out if it ever came to pass.

Don't panic, though: I can't see that we're any closer to achieving it than we were 50 years ago.

A computer did indeed become world chess champion. IBM's Deep Blue beat Garry Kasparov in 1997, so Newell and Simon were almost 30 years out in their 10-year prediction. By that time, most researchers had given up working on one machine that could do everything, and broken down general intelligence into more manageable problems.

Many AI researchers will tell you what you mainly learn by trying to build machines that think like a human is just how many different types of thinking a human being can do.

Imagine the first hour of your typical day.

If, like me, you're not a morning person, a lot of what you do is performed on autopilot. I'm not really conscious that I have showered, made a cup of tea, put on my pants before my trousers,* and so on. Those are now habits, automatic sequences of actions. I don't even need a sign on the hall door. Nevertheless, I can still do them if circumstances change. If my flatmate's in the shower before me, I can change the order of tasks and make tea first. I can wash up a mug if there aren't any clean.

---

* For our American readers, put on my briefs before my pants.

For a machine, simply telling the difference between a mug and a milk carton can be a problem, let alone deciding if it's clean. Knowing what is happening in the world, and making a decision about changing the order of tasks is at least two problems. Being able to pour tea AND climb stairs is a combination of motor skills beyond most robots. Even a robot waterproof enough to survive 10 minutes in the shower.

And that's just the routine stuff. At the same time I am listening to the radio, composing brilliant arguments against whoever is on the Today* programme that I may possibly tweet but more likely will just shout at the radio. Then I have to read the emotions in the face of my flatmate who got out of the shower while I was shouting at the radio, and possibly apologise for startling him.

In parallel I'm remembering what I have to do that day, weighing up how likely it is that the cobbler will have my shoes ready, pondering whether it would be worth having children just to run errands for me, and feeling a pang of gratitude to whoever invented central heating so I don't have to light a coal fire before I start work.

Nobody has yet managed to instil feelings of gratitude, or of any other emotion, in a machine. And just getting one AI to switch between two types of task with anything approaching the fluency of a human being is still a monumental task. So I am not one of those worrying about the singularity and the triumph of the super-intelligent robots.

But I have met a number of very smart people who told me not only that a machine with super-human intelligence was possible, but that it was already here. And though I don't entirely believe them … well, imagine that the screen's gone wiggly and you're hearing *going back in time* music …

---

* For non-British readers, it's a morning news programme on which politicians and other important people are interviewed about the burning issues of the day. The BBC's commitment to balance on political issues means it's almost guaranteed to annoy you at least once, whatever your political leanings.

The singularity

It's August 2014. I've just landed at Los Angeles airport, and I get a message from a BBC radio producer. They want me to present a documentary about the singularity, which is great news. It's also amazing timing, as I'm on my way to Silicon Valley for a fortnight. If the singularity is happening anywhere, it's there. Half the people we want to interview are in San Francisco and the Bay Area. And, on a whim, I've packed my radio recording kit.

When I Skype the producer, I joke that this lucky coincidence might in fact be evidence that the singularity is already here. Perhaps it's a super-human machine intelligence that organised my presence in California at exactly the right time. I didn't know that I'd be presenting the programme when I booked my flights, or packed my bag, or boarded the plane. Even the producer probably didn't know. But let's suppose there's a super-intelligent network of computers with access to the internet. It would know, from radio listings and my Twitter feed, that I'd worked with this producer before, and done public events on robotics and AI. If it also has access to BBC emails it would know this programme was in the commissioning pipeline. Putting that together with the other data at its digital disposal, it could calculate the probability that, at some point in August, I'd be asked to present exactly this programme.

I realise this sounds crazy, and I'm not saying it's true. But bear with me.

I was in California mainly for an informal weekend called Science Foo that gathers a bunch of interesting people in Google HQ in Silicon Valley: artists, mathematicians, historians, physicists, roboticists, AI people … So I spent a lot of time sticking a microphone under people's noses and asking them whether they thought the singularity would ever happen.

Lots of them thought it would. Some of them thought it already had. Several of them thought it already had, and we were in it. It was a bit disconcerting to sit in Google's canteen and hear historian and technology expert George Dyson[*] declare, 'This *is* the singularity.'

He thought it was very ironic that we'd just been sitting around one of Google's cosy meeting rooms, well supplied with soft drinks and snacks, talking about how far in the future the singularity might lie.

'To my view, there's a very good chance, if not good evidence, that it's happened already. And the fact that we have no proof that it happened, in a paradoxical way, is proof that it is happening', said Dyson.

'Because a real artificial intelligence would be smart enough to not reveal itself. It would just quietly hire people like this who work here and keep it growing and well fed, and the people would be very well fed and paid to take care of this growing AI. To me, when I look at what's happening here, that's what I see.'

So all the people who work here are just subsidiaries to the artificial intelligence that is Google, that is running the world?

'Well, yes, they're helping it grow. Which is not necessarily a good or a bad thing, it's the way the world is', he responds.

'If you imagine a world with a real AI, this is exactly what it would do, it would surround itself with very happy, healthy people who write code and build networks and are thinking about self-driving cars and all the things that are going on here. Why people expect the singularity would be some apocalyptic thing that would suddenly announce itself is just silly.'

Instead, it's plying us with lovely dinners and nice glasses of Shiraz and Pinot Noir …

'Exactly. It would be a very stupid AI that tried to force people to do things they didn't want to do.'

---

[*] He wrote about this stuff back in the twentieth century, in his book *Darwin Among the Machines*, named after the Samuel Butler essay.

So when I joked about Google having planned the whole thing, having used its clever algorithms to calculate that there was a reasonable chance the radio programme would be commissioned, and that I'd be working on it, and that's why it invited me to come here, so I could interview people in its own canteen, perhaps I wasn't so far off.

Which would also mean that it knew everything we said in the interview. So it knows we know about it.

No, I still don't actually believe that Google is running the world. But I had one last coincidence on the way home.

My flight back to London was very full, and by the time I checked in online, I had a choice of about three seats. I went for the best of a bad lot, an aisle seat with only one passenger between me and the window. And what a travelling companion he turned out to be. Californian Jeff Newkirk had so many stories that it seemed a waste to sleep at all. For example:

> *My mom was an early IBMer. She worked for a gentleman by the name of Jack Bertram, who was in charge of research at the time, in San Jose California. This must have been in the early 1960s. They were working on a computer that would speak, and they worked for months on getting the computer to say: "Good morning Mr Watson." This was T. J. Watson who was then president of IBM, he was coming out from New York to visit.*
>
> *They accomplished their task, and about a week before his arrival they got a call to say that Mr Watson wouldn't be there until the afternoon. So it took them one full week, almost 24 hours a day, to change the computer to say: "Good afternoon Mr Watson." That was a really big deal at the time, it was huge.*

Now, you may say that, flying back from California, I shouldn't be surprised about sitting next to somebody with family connections to the computer industry, and I agree with you. But if you *were* a super-intelligent computer with access to the internet, airline seats would be one of the easier things to monitor, predict and even tinker with.

I bought it online

It hasn't taken long for online to become the normal way to do things. I bought my flight online, I entered my passenger details online to satisfy security regulations, and I checked in online. Nobody got paid to turn my information into digital form, I did it myself by typing into boxes on the airline's website.

Internet shopping celebrated its 20th birthday in 2014. On August 11 1994, Phil Brandenburger of Philadelphia bought a Sting album from a small company called Net Market. Their encryption technology meant he could send his credit card details securely from his computer to Net Market's without any eavesdropper being able to steal it. Not even the NSA, the American government's codebreaking equivalent to GCHQ, could decode the communication.

If you're under 30, it may be hard for you to imagine the world of 1994. The *New York Times* reported Mr Brandenburger's purchase, explaining that he visited Net Market's 'store front' in a 'service of the internet called the World Wide Web'. It also noted that Net Market already sold things, but that this was the first secure transaction.

And, although the fledgling encryption techniques in use at the time all relied on the same RSA* mathematical techniques, Net Market had chosen a version that was secure from both criminals and the NSA. The government-approved version was secure from everybody except the NSA, who would hold the equivalent of a spare set of keys to everything you might do online.

The government agencies who once pioneered computer technology trying to decode enemy radio transmissions now pit their wits against the encryption techniques used

---

* Named after its three inventors, Rivest, Shamir and Adleman, it uses multiplication of very large prime numbers to produce an encoded message that can be decoded by sender or recipient, but not by a third party.

by individuals for private emails, web browsing and online purchases. You may feel that if you have nothing to hide, you have nothing to fear. In which case, congratulations for never feeling embarrassed about anything you said, searched for or bought online.

You could argue that online shopping was already 10 years old when Mr Brandenburger made his choice to be perpetually known as the man who bought *Ten Summoner's Tales*. And that the first purchase was not a CD, but cornflakes and eggs.

Jane Snowball, aged 72, was recovering from a hip operation at her home in the north-east of England when she agreed to be part of the Gateshead Shopping Experiment. The local council worked with tech company Videotex to provide home shopping services through people's television sets, sending orders down the telephone line using a special remote control. In May 1984, she sent her first order to Tesco supermarket.

There were drawbacks. Credit cards were not common in 1984 in the UK, so Mrs Snowball had to pay cash on delivery. She chose her shopping from a text list on her television screen, typing in numbers for each item. And she missed the human interaction of shopping. In many ways Michael Aldrich of Videotex was too far ahead of his time.

Today, we can choose our online purchases by clicking on pictures of eggs or cornflakes. We can listen to extracts of the Sting album before we buy it. And if we do choose to visit a real supermarket, we'll probably be forced to interact with a machine anyway.

Online shopping has many advantages for the consumer. You can do it when it suits you, which often seems to be on work time. The day after a weekend or bank holiday are peak times for online shopping. You can quickly check out competing offers without having to walk from shop to shop, and you don't have to carry your own shopping home: especially handy if you've just bought a fridge. You don't even have to get dressed, unless you're doing the sneaky-shop-at-work thing, in which case shopping in your underpants might draw too much attention.

And online retailers are often cheaper, probably because they don't have to run actual shops, which can be expensive. Especially when shoppers come in, look at products in your showroom, ask your staff lots of questions, and then go home and buy the same thing online from your competitor.

Most of these things are also advantages for the retailer. They save money on having shops, and on transport as the stock can stay in the warehouse till they deliver it to your house. They can take your money at any time without paying a person to be nice to you and patient with your unreasonable demands to try on the same thing in three colours and seven sizes. I'm not sure they benefit directly from you shopping in your underpants, unless they're a company selling underpants.

But unlike a real shopfront on a real street, an internet shopfront can't rely on you wandering past. They have to find other ways of saying, 'hey, we're over here!' to potential customers. Luckily for them, the virtual High Street has a few features they can use.

Imagine that, wandering down the real High Street, you left a visible trace that showed not only which shop windows you looked into, but also how long you stayed and what you looked at. Imagine your conversations, not only with shop assistants, but even with your companions, were recorded and played back by somebody picking out key words such as 'shoes', 'expensive' or 'wedding'. Imagine that this information was being combined with other things nobody could tell just by looking at you: your age, your postcode, maybe your sexual preferences.

Now imagine that the shop assistants are SO keen to sell to you that they rearrange their window displays before you pass. They've fed all this information into a computer, which predicts what's most likely to appeal to you and how much you're likely to pay for it.

That, in effect, is what's happening when you shop online. Not only your previous shopping history, but other information such as your online searches, what you post on Facebook and Twitter, and who your friends are, can be used to target you with adverts, and affect which version of a website you see.

Which is why you get adverts for things you've browsed online and possibly already bought. Or why you may be offered a different price for an online purchase to what your friends are offered, even by the same seller.

And if you're thinking that you'll go back to physical shopping, where you are safely anonymous and the shop assistants are too busy to rearrange their window displays for each new customer, don't feel too smug. You're giving out more data than you probably realise, even walking down a real High Street.

If you have a cellphone that can do more than just make and receive phone calls, it's collecting and sharing all sorts of data about you. Those apps that track your running know where you've been and how long it took you to get there. If your cellphone has plugged into somebody else's Wi-Fi internet, or even searched around for a potential connection, it's exchanged little packets of data that identify you and quite likely your home postcode.*

And remember that the joy of big data is its ability to connect different databases to see a bigger picture. If you've also used an Oyster card or other travel card, and paid for a purchase with a bank card, you're on three different databases. Even if they're not combined in a way that identifies you individually, pooling the records gives a general snapshot of that day's shoppers: How far have they come? How much have they spent? What are their home postcodes?

We all leave a trail of digital breadcrumbs, or digital exhaust, which can be useful to all sorts of people, not just the ones trying to sell you stuff. Health services find it handy to know who is searching for advice on flu symptoms. City planners find it useful to know what journeys we make.

---

* Or zip code, if you're American. Here in Britain, a zip is what holds up your trousers. Or pants, as you call them.

Talking your language

Using all this data takes a combination of machine and human intelligence.

Machines have come a long way since the IBM computer took a week to learn to say 'Good afternoon' to Mr Watson. Now IBM has a computer called Watson that can understand human language. Watson proved this, not by passing a Turing Test but by winning a US television game show called Jeopardy against two human competitors.

This apparently frivolous achievement was a demonstration of Watson's ability to combine several human thought processes. First, it had to understand the questions, which in Jeopardy are a sneaky mix of puns and obscure references, wrapped in a slightly odd grammatical structure: the quizmaster gives an answer, or clue, and you have to provide the question.

Next, Watson had to search through all the things it had learned, or at least stored in its memory, for the most likely solution. In many cases, it wouldn't know the answer for sure, but could make a reasonable guess. Then, it had to construct an answer in the correct form, decide how much to bet on it being right, press its buzzer and deliver the answer out loud before one of the other competitors could get there.

Paul Horn was director of IBM Research when he proposed that the company's next Grand Challenge should be a machine that could pass Turing's Test. A previous Grand Challenge had given them Deep Blue, the computer that beat Garry Kasparov at chess. Could the next one fool a person into believing it was human?

By now, it was clear that natural language, the way real people talk and write to one another, was one of the hardest things for computers to learn. Not only is human language governed by complex systems of logical rules that we call grammar, but languages are also riddled with exceptions and inconsistent pronunciation.

English uses words and syntax from many different roots, including French, German, Celtic and Scandinavian languages, vocabulary from the Indian subcontinent and Arabic. It contains dialects that only became welded into one language after writing became widespread, and which make it hard for a human non-local to understand exchanges such as: 'Fit like?' 'Nae bad, fit like yersen?' 'Charvin'!*

If written language is tough enough, spoken language is a whole new layer of pain for a computer. Any very odd typos in this book should be blamed on the dictation software I used, which shows scant understanding of the difference between sense and nonsense. Or scent sand nonce ants, as it would say.

Human language depends on context. You are constantly learning new bits of vocabulary, or even structure, by reading and listening. And, though you may occasionally look up a word in a dictionary, mostly you work out what it means from how it's being used. What's more, you recognise that its meaning might be different in a different setting. 'Tying the knot' means one thing in a marriage registry office, and something entirely different in a vasectomy clinic.

So it's a tall order for a machine to cope with language the same way that we do. But it's also crucial, if artificial intelligence is to be available to everybody, not just the programmers and coders who can translate human thoughts into digital data.

That's why Paul Horn thought that Jeopardy would be such a good test. A machine that could understand and answer the convoluted clues, under time pressure, might not be passing the Turing test, but it would be closer to doing so than ever before. And, unlike the Turing Test, it would draw an audience of millions on television.

---

* For any computers, or non-Aberdonians, reading this: 'How are you?' 'Not bad, how are you?' 'Absolutely marvellous!'

The IBM research department was sceptical at first, both that it was a worthwhile task, and that it was even possible. But a few people put together some software that might be up to the task and started training it with old Jeopardy questions. Like its human rivals, Watson would not be allowed to use the internet to help.

If, like me, you're on a pub quiz* team, you know that most answers are not 100 per cent certain. There's usually some argument between team members who are more or less sure they know how many countries have borders with Switzerland. But you have to decide which is the best shot, according to such scientific criteria as who reads the international news, who has been there on holiday, and who has a vastly inflated idea of their own general knowledge.

Watson does all this internally, by finding different possible answers and weighting each one according to how reliable, relevant and generally right it's likely to be. This process draws on what Watson learned from previous questions and answers. It's not infallible, but it has the advantage of speed over the human competitors.

So, though Watson made a few terrible errors, such as calling Toronto a US city, it beat reigning Jeopardy champions Brad Rutter and Ken Jennings to the $1 million prize in February 2011. Watson's cooling fans were deafening, and it took up an entire room, so it was represented in the TV studio by an illuminated logo and a synthetic voice.

Nobody was fooled into thinking Watson was a human being, but I think Turing would have been impressed.

Watson has not milked that triumph by touring the world's gameshows. It took dozens of researchers over five years to get their protégé to the winning position, and perhaps they felt it was time for Watson to grow up and get a proper job.

---

* Or a trivia night, as Americans call them. See? Another regional variation.

So IBM's Watson division, founded in August 2011, sent their robot child to medical school.

Doctors have to weigh up information against their medical knowledge, ask the right questions of patients, look for the most likely diagnosis and choose the best option for treatment. IBM, noting that medical information doubles every three years, want Watson to be the ideal physician's assistant, one with time to read all the new research papers on obscure diseases, analyse the hospital lab results, and be able to learn and to come up with new hypotheses.

These days, Watson is allowed to connect to the internet.

Like your annoyingly overachieving mate from school, Watson isn't just fighting cancer and making health care more efficient. For more on what IBM Watson is doing, see the extra chapter of updates at the end of the book.

And Watson is ambitious, with an eye on the White House. IBM suggest that Watson's cognitive computing can help governments provide better services, respond to their citizens' needs, and engage better with a disengaged public. And, in a less cuddly role, keep an eye out for abnormal behaviour that could signal a security problem.

Soon, Watson could be in your pocket. Three technology companies shared IBM's latest grand prize by developing mobile apps that will fit Watson's data-analysing talents into your cellphone. GenieMD is a healthcare app for patients and their families; Red Ant's Sell Smart app helps retail assistants to sell the customer what they want; and Majestyk Apps made FANG, a soft toy that can talk to a child, answer questions and even ask questions in return.

IBM's Watson is not the only thinking machine that aspires to connect every individual to the world of big data via a portable device, of course. Apple's SIRI, Microsoft's Cortana and Google Now are all designed to answer or pre-empt your needs by combining what they learn about you, individually, with data about entire populations of people who are, in some way, like you.

If you hate the vision of the modern family ignoring each other at the dinner table as they look at their own individual smartphone, tablet or plush toy, you might prefer Jibo. Jibo is a 'social robot' created by Cynthia Breazeal, of MIT Media Lab's Personal Robots Group.

Jibo will answer your questions, anticipate your needs and remind you to check at the cobblers to see if your shoes are ready. But, unlike your cellphone, it will do the same for the rest of the household. The prototype can recognise faces, greet people by name and move its 'head' to suggest it's paying attention to you.

It can't yet mix you a cocktail, but it probably can say, 'you look like you've had a hard day! Here, let me suggest some cocktail recipes …' based on what it knows about you, your drinking habits, and what's in your fridge.

Soon your fridge will be able to do the same thing. Smart fridges can already send a picture of their contents to your cellphone, like some kind of weird internal selfie. LG fridges can engage in text chat about important things, such as how many beers you have left.

The Internet Of Things means that every electrical device will soon be online. While you're out, your toaster can look at provocative videos of bread rising in an oven. Your washing machine can gossip with the neighbours' laundry equipment about who has the dirtiest towels. And your electricity supplier will know exactly when your central heating comes on, how long you spend in the shower, and how late your teenagers stayed up watching TV.

OK, I made up the stuff about the toaster and the laundry. But the other bit is true. By collecting that kind of information from every household, businesses and government departments will be able to plan how much electricity the population will need, when. Longer term, they may be able to reduce wasted energy, by manufacturing goods that better fit how real people live, or by getting us to change our behaviour to fit their goals.

The big data era is already here. Almost everything you do in modern life spits out digital data, and the smart machines, and smarter methods, for making sense of it are getting faster and cleverer every year.

So, what has big data ever done for us?

PART 2: WHAT HAS BIG DATA EVER DONE FOR US?

Fifty-seven notches on a wolf bone translate to 111001 in binary code. That's just six bits* of information. Eight bits of information equals one byte. So one wolf bone is less than one byte.

My laptop can store 120 gigabytes of information, equivalent to over 120 billion wolf bones, but more practical. I have trouble finding anything in my office already, without sorting through billions of bones for the one I need.

However, it's not just the quantity of virtual notches that holds the promise of big data. There's the multiple dimensions, mixing wolf shin bones with mammoth mandibles, shrew femurs, and even moth wings. There's the way all the bones automatically fell into the cave, ready-marked, without a human having to raise a flint. There's the speed of collecting and sorting the bones, and the ease of predicting next year's wolf population.

So, before we go on to explore all the things big data does for us, and promises to do in the future, here's a question for you to consider: Why, instead of big data, don't we talk about automatic data, or timely data, or multidimensional data?

Perhaps none of those phrases hold enough blockbuster appeal. Say big data aloud. Go on, nobody's listening.†

---

* From 'binary digit' – which can be zero or one.
† Oops, nobody except that man reading the newspaper on the seat behind. Sorry about that.

Done it? I bet you said it in a movie-trailer voice, didn't you? 'BIG DATA' in the deepest voice you could muster. I've been talking about it for years, and I've only just managed to say it in a normal voice.

'The name's Data. BIG Data.'

It sounds impressive, powerful, it trumps whatever came before, which was just data, or statistics, or notches. It's like getting your dad involved in your playground fight. 'My data's bigger than your data.'

Don't get me wrong: it's snappy, it's memorable, and it fits on the cover of a book.

But be careful, when big data is wooing you with its seductive claims of being *very* objective and *so* scientific and irresistibly *rigorous*, that you're not falling for the implied omnipotence of its powerful computers and its *huge* sets of data.

Remember, size isn't everything.

CHAPTER FOUR

Big business

Before I was born, my mother worked for the Gas* Board in the north-east of England. She didn't use a computer. She almost was a computer, in the old-fashioned sense of a woman with a pen and paper, entering figures on to a spreadsheet. A paper spreadsheet. Her job was to work her way through a pile of gas bills, copying numbers on to big sheets of paper. After weeks of work she took the results into her manager's office next door.

'Oh, those?' he said blithely, 'we've already estimated them.' And he threw her handiwork straight into the bin, in front of her. They couldn't wait for the actual figures to be painstakingly collected and transcribed, so they'd made an educated guess. I don't think she was particularly offended, but she remembered it as a ludicrous waste of human time.

Today, the Gas Board's successor, British Gas, is gradually introducing one integrated system that uses smart metering, the direct collection of digital meter readings in real time. So the manager can call up the information he needs instantly, because it is already on his own computer. And my mum would be out of a job, or she'd be a data analyst or something.

Not only that, but everyone who has a smart meter installed can see the patterns of their energy use, and have an app on their mobile phones to control their heating and hot water, because the boiler is connected to the internet. So even if you live alone, you can send a message from the train to get the heating on ready for your return.

---

* For American readers, this is gas that's burned in boilers and domestic heaters, not the stuff you call gasoline and put in your car.

In fact, when the app detects that you're nearly home, it will send *you* a message to suggest you turn the heating on.

From drilling to billing, the energy industry is now using big data techniques. It's the ideal candidate, dealing with large quantities of something that is highly quantifiable and also highly valuable, so it's worth investing in technology that could increase productivity. The combination of large scale and high value means that even small, incremental improvements can translate into millions, or billions, of dollars saved.

The digital oil field has been around for years, connecting incoming data from sensors built into drilling and pumping equipment to provide a virtual model of what's happening, where oil is flowing and how fast, and which parts are not functioning as they should. If one of the sensors suddenly shows a drop in pressure, that could be a leak. Knowing about it within minutes, instead of waiting until the next inspection, lets them repair it days or weeks earlier – saving both money and a damaging and expensive clean-up job.

A network of automatic devices in direct communication, the Internet Of Things, can monitor safety as well as wear and tear. The advantages of automatic collection and relaying of information, and the feedback systems that can address a problem without having to wait for a human being, mean oil and gas were among the earliest big data industries.

As the technology for collecting and processing the data gets faster, cheaper and more accurate, oil engineers can go beyond monitoring past and present problems, to predict future problems. Knowing instantly that a valve has failed is better than waiting till there's a visible leak, or worse. But identifying the risk that a valve will fail within the next week, or day, or hour, means they can fix or replace it before the problem arrives.

Predicting when your car will need a new oil filter, so the repair workshop can have it in stock ready for you, might save you a wasted day. Getting a replacement part to an oil drilling rig off the Nigerian coast, however, can take four months.

That's a long, costly wait for a lot of highly skilled professionals and expensive equipment.

What's more, by connecting the engineering side of maintenance with information about logistics, transport and personnel, everything can be made more efficient. No more sending a specialist halfway around the world to do one job, or keeping ships hanging about between trips. It costs tens of thousands of dollars to keep a large oil tanker at sea for a day, so even small improvements in efficiency can save hundreds of thousands of dollars.

Transportation in general benefits from the kind of detail and near-instant updating that big data offers. In the aviation industry, for example, fuel makes up around a quarter of all costs. Saving 1 per cent of the fuel used on every route, as GE claim to have done for Air Asia by using big data analysis, quickly adds up to millions of dollars.

Rolls-Royce build sensors into their aircraft engines that can notify maintenance crews when servicing is needed to avert future problems. They call their system Engine Health Management.

Basic readings such as oil pressure, fuel flow, temperature and speed can be read by the pilot, but selected data can also be transferred to remote control centres, via a wireless internet connection at the destination airport, or relayed via satellites while the plane is still in flight.

Detectors can read the magnetic traces of metallic particles inside the engine, or changes in vibration, and use information such as airspeed to build up a multidimensional picture of what the engine is doing. Then, abnormal patterns of engine behaviour are picked up automatically by software, so problems can be foreseen and potentially dangerous mid-air failures can be avoided. The monitoring service can liaise with the aircraft operator to arrange inspection by a human engineer, with minimum interruption of the flight schedule.

Modelling aerodynamics, the performance of jet engines and the weather along tomorrow's route are hard problems. But they look simple next to predicting how many people will want to fly a route in six months' time, how much luggage they will bring, or how many potential customers you will lose when you overbook a flight and turn away two journalists with Twitter accounts.[*] But many big data companies promise exactly that: better understanding of human behaviour by analysing the data trail we all leave behind.

Model customers

The loan company Wonga has been the subject of some controversy in the UK. They specialise in short-term loans, repaid over a few weeks, at high rates of interest. They make no secret of the cost, either the rate of interest or the extra charges. The example on their website is a loan charged at 1,509 per cent APR, at least 100 times the rate my bank would charge me, and around 500 times what I'd pay on a so-called 0 per cent credit card transfer.[†]

New rules from the Financial Conduct Authority (FCA) capped the maximum daily interest rate[‡] and introduced tighter affordability criteria. The company has also changed its marketing approach and written off some loans to people who wouldn't have qualified under the new rules. They're still a lot more expensive than my bank or credit card, though they're also more upfront about the charges.

Why would anybody go to them? Wonga offer a quick decision, from the comfort of your own computer or smartphone. Money can be transferred into your bank account within minutes of you filling in an online application

---

[*] Which may seem like a far-fetched example, but since I was one of those journalists refused boarding on an overbooked Easyjet flight, forcing me to delay the start of a performance in Belfast and keep 150 people waiting for half an hour, it's only too real for me.

[†] Because there's always a 3 per cent admin fee, isn't there?

[‡] Wonga's interest rates used to be over 5,000 per cent APR.

form. So it's fast, anonymous, and you can do it in your underpants. But also, because you don't have other options.

Banks and credit cards don't like lending to people who can't repay them, or who will be expensive and time-consuming to pursue. So if you don't have a good track record of borrowing and repaying money, it can be hard to convince a major financial institution to lend you money now. Ironically, all my years as a feckless freelancer, borrowing money in thin times and repaying when my invoices get paid, seem to have given me a decent credit rating.

But even before the new, stricter regime, company founder Errol Damelin boasted that Wonga rejected 60 per cent of loan applicants and had a default rate of 7 per cent, which is below the usual 10 per cent default rate on credit card lending. Given that Wonga's customers tend to be more strapped for cash than the average bank customer, how did they do it?

When Damelin started his first loan company, samedaycash, he had a default rate of 50 per cent. Going by income, that's not a good business model. But Damelin wasn't interested in collecting money, at that stage. He was collecting data. With so much information readily available about individuals, why stick to the kind of credit history that banks were buying in from credit agencies, or previous borrowing records?

In an interview with *Wired* in 2011, South African Damelin took his faith in data on to an ideological basis. 'Prejudice and generalisation are something I grew up with,' he said. 'I think when people are saying that a good old bank manager should make the decisions, what they're really saying is some middle-aged white guy should make the decisions.'

Instead, Wonga collects a few dozen pieces of information from a potential customer, and uses those to gather literally thousands of other bits of data. The artificial intelligence in Wonga's system uses all that information to decide whether to lend you money.

What kind of data? Well, for a start, Wonga knows what software and hardware you're using to apply for this loan, and where you are, geographically. It's also keen to hook up with you on Facebook, where it will get access to your friends, to lists of what you like … all sorts of details that an old-fashioned bank manager would never know.

Unlike the 'middle-aged white guy', Wonga's computer makes no moral judgements about your choice of friends, music, partying behaviour or cat videos. It only wants to know whether the thousands of dimensions of your life correlate with you being likely to repay a loan. No individual human being is going to trawl through this intimate snapshot of your life, but a computer will, and that computer decides whether or not to loan you money till pay day.

The kind of technology that Wonga uses, especially through Facebook, is not very different from the methods that target you with online adverts for things you might be interested in. Instead of just going on your own behaviour and expressed tastes, however, it's also using what it knows about your friends and acquaintances. If they're all good borrowers who repay on time, chances are that you are too.

If you're like me*, your mind is already wondering why: is it because you can borrow off a mate if you get stuck, or because you probably know them through work, which means you all have decent incomes, or simply that responsible people tend not to stay friends with reckless idiots who squander all their cash in the week after pay day and then have to borrow money to pay the bills?

I don't know. Nor does Wonga. All they know is that they can build up a multidimensional model that gives them the odds you'll pay up on time, or a week later with an added late repayment fee. They don't need to know why, they're just looking for patterns.

How you feel about this new type of credit rating may depend on who and where you are. I prefer to borrow money

---

* Since you're reading this book, there's an above-average chance that you are like me, on some relevant measures at least. Or somebody bought it for you, and you're very polite indeed.

from financial institutions with whom I have a purely business relationship. If I lived in India or parts of Africa, I might not have that option. I might not even have a bank account. But I probably would have a cellphone, and my record of using and paying for that could give potential lenders clues about my likelihood of repaying a debt.

Already, many thousands of loans have been offered to people, from Colombia to the Philippines, based on measures like social media activity. Some of these loans are for emergency expenses, but many are for small businesses to invest in the stock or equipment they need to grow. Lack of access to banking services, including credit, is often seen as something that holds back economic progress, or even as a form of social exclusion. Being able to use your social media history, instead of just financial information, to access that credit is apparently a welcome option for many.

It's not surprising that people lending you money want to get to know you, to form an idea of how likely you are to repay them. But companies selling you products also have a vested interest in predicting your future behaviour. Nobody wants to be left with a supermarket full of unsold food. So analysing consumer tastes is big business in itself.

Consumer intelligence

What did you mostly eat in 2015? Any changes to your shopping and dining patterns? Did you, for example, find yourself looking for the words 'responsibly produced' on the packaging, or 'natural sweeteners'? Were you using more products produced in small batches, or to religious standards? Did you develop a taste for fermented foods?

Those were the five trends predicted for 2015 by dunnhumby USA, who call themselves a customer science company.

What distinguishes a customer science company from old-fashioned market research? Any supermarket can notice that sauerkraut and drinking yoghurt are selling faster, read some trendy restaurant reviews and take a gamble that kimchi[*] and kefir[†] will be the next big thing. And getting some of the customers together to say why they're buying things isn't new either – market researchers and opinion pollsters have run focus groups since the 1950s, adopting a method developed for social science research in the 1920s. Why were dunnhumby USA so confident in their predictions?

For a start, the dunnhumby group has access to 770 million shoppers around the world. And by access, they don't just mean that's how many customers they could potentially stop with a questionnaire as they leave the store, they mean they have already collected data from these people.

When I was little, my mother would give me her Green Shield Stamps collecting book and a pile of the stamps, and leave me happily tearing along the perforations, licking and sticking them into the book. Twenty completed books could be exchanged for a compact camera, while a portable personal cassette player[‡] could be yours for a mere two books.

I have no idea how much she had to spend to get each stamp, though I do remember them being handed over at the till on our weekly supermarket visits. I don't remember us ever claiming anything from the catalogue, though, so I suspect their main function was to keep me quiet for half an hour.

Green Shield Stamps disappeared in 1991.[§] In 1995, Tesco supermarkets launched their own loyalty scheme, offering shoppers rewards that they could spend, in Tesco shops of course, collected on a Tesco Clubcard.

There are a few advantages to this kind of loyalty scheme. It's harder to acquire fake Clubcard points than to forge green

---

[*] Korean pickled-cabbage dish with chilli, like sauerkraut went away and learned karate. During 2015, it was one of the year's fashionable foods, in London and New York at least.

[†] Like yoghurt, but more authentic.

[‡] Ask your mum.

[§] When their catalogue stores became Argos catalogue shops.

Shield Stamps, which people did. You're not offering rewards that people will spend in a different shop. And it's much quicker for your staff to swipe a plastic card than to flick through a paper book in which some snotty child has licked and stuck hundreds of paper stamps. More hygienic, too.

But the important difference is that, unlike paper stamps, a plastic card can collect information for you about what was bought, when, and in what combinations.

The Clubcard scheme is credited with putting Tesco ahead of main rivals Sainsbury's, by attracting customers with points they can spend in only one store, but also by enabling the store to understand and predict shoppers' behaviour. And to help them with this new project, Tesco hired a couple working out of their spare bedroom, Edwina Dunn and Clive Humby.

When Clive left his previous job to start a business, his employers sacked his wife, Edwina, fearing a conflict of loyalty. That may not have been very ethical, but by doing so they created a near-unstoppable force in customer data collection and analysis. Clive and Edwina's company, dunnhumby, ran an initial trial for Tesco, whose chief executive is reported to have sat in stunned silence when they reported back, and then told them, 'You know more about my customers after three months than I know after 30 years.'

Why does this matter?

An often-repeated story about early data-mining reports that customers tended to buy beer and diapers, or nappies as we call them in the UK, together between 5pm and 7pm.

So in principle, though there's no evidence the store chain in question ever did this, you could move the nappies closer to the beer, or vice versa. Or you might decide that making sure they were at opposite ends of the store will force your thirsty parents to walk past lots of other products, some of which they will buy in their sleep-deprived state of new parenthood.

This case pre-dates store cards, so the data analyst who found the unexpected beer-and-diapers relationship worked from the checkout data. All they had was the list of what was in each shopper's basket. There was no way of knowing anything else about those shoppers. Some guessed that they were young fathers buying something for the baby and something for themselves, possibly forgetting that women also drink beer.

Today, Tesco can send targeted offers to Clubcard holders, so everyone who buys diapers more than once could get money off beer. But that's just the start. Buying diapers now means you'll probably be buying baby food soon, then children's clothes, then school uniforms.

By putting that together with your other buying patterns, the store can profile you. Have you chosen the small-batch, organic beer? Then you may move towards a more health-conscious diet as your children grow up, more natural sweeteners and fewer ready meals.

Combining your data with other local information, they might predict that you'll move house within the next five years, as you need more space for your family. Or that you'll stop using their local store and start driving with the kids to the big supermarket down the road instead. Or that your child's mother, or father, will divorce you for your beer-drinking ways and you'll be buying ready meals for one, extravagant toys at half-term and Christmas, and way too much alcohol.

There's no guarantee you, as an individual, will follow any of these trends, but for a supermarket selling millions of products every day, that doesn't matter. It's percentage points that count. Instead of advertising to a population that's only 10 per cent likely to buy small-batch, naturally sweetened, kosher* kimchi, they can focus on the population that's 78 per cent likely to buy it.

---

* If you're thinking that only observant practitioners of certain religions buy religious-standard food, dunnhumby say you're wrong. Apparently, other shoppers are attracted by the idea that halal foods, for example, are more natural and less processed.

Clive Humby and Edwina Dunn estimated that they got over £90 million when they sold their share in dunnhumby to Tesco. Perhaps they foresaw that Clubcards would become redundant. Today, it's easy to put together data from the bank cards we use to pay for our shopping, the cellphones that constantly report where we are, and even what we say on social media about our plans for the weekend, and use big data techniques to profile us and predict what we'll do in future.

So Humby & Dunn, as Clive and Edwina now have to call themselves,* have moved on. Their new projects take big data into the world of social media, building mass relationships with fans via Starcount, and developing audiences for arts and entertainment with Purple Seven. They don't need you to carry a store card any more.

Enough of you is out there to piece together your loves, hates, habits and desires.

Ninety-nine per cent intuition

Silicon Valley spreads out from Stanford University, alma mater of so many millionaire tech company founders, around the south-east end of San Francisco Bay. Google's campus, Facebook's headquarters, and even the Museum of Computing are here, along with countless smaller companies and outposts of bigger organisations such as NASA. The quiet streets, lined with trees to shade pedestrians from the Californian sun, contrast with busy San Francisco, only a few miles away at the end of the Bay.

Unusually in America, the Bay Area has a comprehensive public transport system, from trams to trains and everything

---

* The perils of naming a company after yourselves and then selling it. If I ever sell my brand, I suppose I'll have to call myself Harkness Timandra. Or I might go for something shorter, like Kim Star. Frankly, if you pay me £93 million, as Clive and Edwina got for dunnhumby, you can call me what you like.

in between. So I climb off the double-decker, air-conditioned CalTrain into the heat of Redwood City station.

The name conjured images of the Wild West in my mind, but they're dashed at once. This is a sleepy little town of pavement cafes and offices. My destination is a 15-minute walk away, and that's enough to take me out of the town centre and into a business park, all drive-in stores and fast-food joints, interspersed with anonymous office blocks.

Interana stands for interactive analytics, and if I were designing the set for a film about a tech company in 2015, it would look like Interana's offices.

When my host arrives, I'm taking a sneaky photograph of Pythagoras' Theorem, spray-painted on to their varnished concrete floor, and chatting to the guy practising his hoverboard skills in the coffee area. I get a quick guided tour of their new home – lots of whiteboards, meeting rooms named things like Space Travel and Radio, and beer for Fridays.

The CEO, however, is not just another young guy in jeans and sneakers. Ann Johnson is elegantly dressed in white linen. A former Intel engineer, she left to set up Interana with her husband, who ran the infrastructure team at Facebook. Now they're expanding so fast that, before they moved in here, their interns had to work sitting on the floor.

We escape from the fun, tech startup vibe into her corner room, more like a Regency salon with its upright sofas and muted grey decor. It's an odd setting for a conversation about Tinder, the dating app that's one of their major clients.

Interana provides a system that lets any Tinder employee use their own company's data to answer questions about their customers' behaviour. 'We get their stream of everything that's happening on their app, from clicks to swipes to messages to performance data,' says Ann.

At this point wouldn't it be great to put in a personal story about how their insights have made my life better, as a Tinder user? But I'm not very photogenic, I'd get far too many 'swipe left for reject' responses for my ego to bear. My dating depends on meeting people face to face and hoping they share enough interest in maths or motorbikes or opera, or find me funny enough, to 'swipe right' in person.

For my friends who do use Tinder, Ann describes how she and her team have made your lives better:

'Tinder has found that the more selective you are, the better matches come out of it. There is a small subset of users in Tinder that were rather indiscriminate, they would swipe right all the time. And I think it was discouraging to some people, who were really hoping that they would find somebody who cared very much about them, specifically.'

Once Tinder had learned this, using Interana's system, they were able to experiment with changing the way the app works, so it rewards more selective behaviour.

'They released in Australia a Superlike feature, where you get one like per day that is a Superlike. Imagine I see you across a smoky room* ... our eyes light up ... that's what I imagine a Superlike is about,' she explains. 'And that came about from getting this intuitive understanding that when people are not choosy enough, the matches aren't great. Then they were able to use human intuition to design a feature around that.'

Johnson is very modest about what her technology can do, and about claims that algorithms can model all the quirks of human behaviour.

'There's no way that Interana could say: You guys should generate a Superlike feature! We just give them the data, so they can make the best product decisions. The idea that you have this amazing machine that you put the data in, and the answers come out, that's what a lot of people imagine happens. It's ... I would say 90 per cent, 99 per cent person, and 1 per cent machine. That person puts hypotheses into the machine and then sees if those are right.'

To illustrate, she tells me a story about a department store that wanted an algorithm to give customer recommendations on their website. They spent so long discussing what kind of

---

* Ask your mum.

algorithm to use that they ran out of time, and said, 'Well you know what, we need it now, so let's just recommend sheets. Let's just always recommend bedsheets.'

Not only did they sell a lot more bedsheets, says Ann, but it took them three years to come up with an algorithm that worked better than simply making a blanket* recommendation of bedsheets.

'Beating human intuition is really hard. I think people underplay ... because technology is so cool, they underestimate what human intuition is.'

But for businesses, the dream is that by understanding what makes your customers tick, you can change what they do. By pulling the subconscious strings of people's minds, you can get them to buy more, or renew their cellphone contract, or recommend you to their friends. So what about moving beyond understanding people's behaviour to *changing* their behaviour?

'The question with big data is always: can you determine causality?' says Ann. 'I think it would be very presumptuous of us to say we can. However, the more informed you are, the better you can approximate causality. So if you can understand how people are behaving to the most minute detail, you can start looking at individual behaviour.

So in that way we can help you circle around the ultimate prize of causality.'

You may not know the why, but you'll have a very detailed picture of the how, and that may get you close enough to try some things out on your customers. Such as recommending bedsheets.

Johnson spends a lot of time, if not looking down the microscope at human behaviour, polishing the lenses to help other people to do so. But, like all of us, she's also one of the people generating data that companies study, and one of the people they try to keep as loyal customers. Did acquiring insider knowledge change the way she feels about being a customer, I ask her? She laughs.

---

* Sorry.

'Yes, that absolutely happened. The reality of how much data everyone has on you, and how they watch you, was staggering. But then very few have incentives to look at you individually, so when I see how much data is out there, we're kind of anonymous in a giant swirl of data.'

She smiles, 'But personally, I'm from a small town in rural Minnesota, and in a small town it is everyone's business to know what you're doing. People don't rob other people's houses because there's other people watching them at all times. If there's a different car parked outside, or if you turn off your light at a different time of night, everyone knows.'

And that's how Johnson sees the data-driven future. 'For me it's a lot like living in rural Minnesota, where you just assume everyone knows everything.'

So the internet's going to be like we're all living in rural Minnesota?

'Exactly. Everyone gets to live in a small town where your business is everyone's business at all times,' she says.

'Weirdly, I like it. You get all these targeted ads now for that thing that you thought you were going to buy, and you looked at it for 10 minutes and didn't buy it. But then you regret not buying it, and it's right there! It's right there in your side bar and you can just purchase it anytime. Or you get great book recommendations based on what you last bought. If my favourite author publishes a new book I'm very confident that I will know about it, because that's a purchase opportunity.'

Like having your personal butler following you around and saying: Hey, you liked that, why not try this?

'Exactly. I don't mind!'

We chat for so long that I'm in danger of missing my train back to San Francisco. One of my hosts suggests an Uber, the suitably big data taxi service you can order from your cellphone.* But, perhaps because I don't come from a small

---

* It uses your location to find the Uber driver closest to you, and you pay via the app.

town in Minnesota, I don't want the Uber app tracking all my movements and using my data, so I don't have it on my cellphone.

Interana calls one for me, and I get my first Uber ride back to the CalTrain station. The driver has lived locally all his life, but he has travelled to Europe, so we compare holiday memories on the brief journey. Turns out, though we've both been to Turkey, Greece and even the volcanic island of Santorini, he has never taken the CalTrain.

I personally recommend it to him.

Butlers off Broadway

That personal butler, following you around with helpful suggestions, is exactly what they're working on in Yahoo! Labs. At 20, Yahoo! is a tribal elder in tech company years, and its labs have been researching for a decade, though they've only recently moved into a new home, just off Broadway and Times Square in New York. It's the former *New York Times* building, and the symbolism of old-fashioned print news displaced by online, data-driven, personalised media is not lost on the residents.

Apart from a couple of table-football tables, there's little effort to make the place look fun or creative. It's a quiet, comfortable, neutral setting in which keen researchers such as Professor Steven Skiena, on sabbatical from Stony Brook University, can work away undisturbed, except by the construction work that rattles Manhattan from dawn to dusk. He gets to work early enough that the cleaners can ask him about the mathematics on the whiteboards. Which is mainly probability, he says.

Back in 1988, Skiena was on a team that won an Apple-sponsored competition to design a computer for the year 2000. It looks very like an iPad. And apart from some grey hairs, he's recognisably the same man as the University of Illinois student who predicted touch-screen typing and wireless communication with other devices.

I'm also welcomed by Head of Global Operations Ken Schmidt, who likens Yahoo!'s early work to a catalogue of web pages by theme: travel, sports, weather. A kind of beginner's guide to the early internet.

Now, when most of us carry a cellphone bearing more computing power than the whole world had when man first walked on the Moon, and access to the countless resources of the World Wide Web, a library catalogue is not enough. We need a personalised guide to show us where to look, says Ken.

This is where Steven Skiena's work with big data comes in. Skiena is excited about the ease with which data is collected today, and the cheapness of storing it for later use.

'Basically any interaction between man and machine, or machine and machine, these days is being logged,' he says. 'You get a tremendous amount of records and there's a lot of interesting things you can learn by looking at them, things that may not be obvious at first, lurking there if you are clever enough to analyse it right. My favourite example is: in the United States there's something called baseball …'

At this point I fear I'm in for yet another session of somebody trying to explain statistics, something I do understand, through the metaphor of sport, which I don't understand at all.

But my fears are groundless. Skiena isn't very interested in sport either, only in the fact that Major League Baseball has player records going back for 150 years.

'Somebody was interested in whether left-handers live shorter lives than right-handers. Now, how would you possibly figure something like this out? You need a record of a large number of people along with their handedness, but you don't know your grandfather's handedness, or your great-grandfather's handedness.'

But the baseball records do tell you about handedness, because:

'It matters whether you're a lefty or a righty.

And so I think that one of the skills that I try to teach my students is to think: How can you take a dataset and repurpose it and do something interesting with it?'

And yet, though his curiosity was fired by the hidden treasure buried in the baseball data, somebody did start off with a research question about left-handedness.

'There's two models. A computer scientist typically will say about the dataset: That's cool, what can we do with this? And other people who have a real problem, let's say social scientists, biologists, have a problem they care about and they're more: Where can we find a dataset that sheds some light on this?'

'You have to bring the two of them together, try to see what you can do and understand the limitations of the dataset for the question you have in mind. Obviously there's no women who played Major League Baseball back then ...'

So you won't learn anything about whether female left-handers lived longer or shorter lives.

Here at Yahoo! Labs, Skiena gets to play with a lot of very large datasets.

'For example, one bit of data we got from the city government was the records of every taxi ride taken in New York over the last three years. This is a good example of big data. Every time you sit in your cab there's a record logged, from when you get in the taxi to when you get out. We had maybe 100 or 200 million records.'

From this data, you can look at travel times, how they vary, how congestion and weather affect journey times, whether some drivers tend to take longer routes than are strictly necessary ...

'One thing we were looking at is: Where are people better tippers? We know where people are picked up, where people were dropped off, you know something about the travel time ...'

So where did people give better tips to taxi drivers?

'We thought people going from the outer boroughs gave bigger tips. Part of the reason I think is that these are physically longer trips. You feel more of a bonding ... this is now non-scientific, this is my speculating. But that's the kind of questions that you can ask, and it is fun to speculate.'

This is one of the limitations of big data. It can describe the world, but it can't necessarily explain it. Skiena is clear about this distinction:

'I think of a theory as an explanation that I can understand: something about the world works like this. And if you believed my theory that people from the outer boroughs drive longer, that's now a theory and an explanation,' he says.

'Complicated models might say that people driving from this location pay this percentage tips, and people from here do this. It's just a raw data thing, it isn't a higher-order explanation.'

Personally, I like to think that people in Queens and Brooklyn are just nicer people. But as I was staying with a friend in Queens I may be biased.

How do you feel? What do you think?

But back to the personal butler. How will it know what you like? Would you prefer a taxi or the subway? Hoboken or Astoria? Coffee or tea? A lot of Skiena's previous work has been about using the large amounts of available data to understand what people are thinking or feeling. Data scientists call this sentiment analysis.

'In my pre-Yahoo! life I did a lot of work on sentiment analysis on social media, things like Twitter feeds, blogs', Skiena says. 'That was trying to directly get at the question: What is it that people are thinking? Can you try to get a sense of this from passively reading the social media feeds? As opposed to conducting a user poll or something, which is a lot more expensive. And the answer is: you certainly can get some kind of insight there.'

It even turns out to have some advantages over just asking people what they think.

'One thing is that you get retroactive data, so I can for example analyse older feeds and older news. Suppose suddenly you want to know: How do people feel about how honest Volkswagen is as a company?'

We're talking in late 2015, when people are not feeling at all good about Volkswagen's honesty.

'You ask people, and right now it's going to be in the toilet. But at some point it was probably very good. If only you could ask them a year ago. If only I could ask them two years ago! Well, if you have access to their social media data, and you can do this kind of sentiment analysis on it, implicitly you can ask them that kind of question.'

This is one of the things Skiena finds powerful. 'You're able to look back at the past. There are surprisingly good digital records going back surprisingly long. The classic example is Google Books. Google has digitised something like 20 per cent of all the books ever published, and they know when these books were published. So you know what people were thinking at the time when they published the book, and you can start to look at, for example, changes in language usage.'

Now, I've had fun with Google Books myself, finding correlations between words and entertaining myself by speculating on reasons for the correlation. Plotting frequency of use of the words 'hamster' and 'contraception', for example, produces a graph in which the two words seem closely linked throughout the twentieth century.

Do hamsters cause contraception? Does contraception cause hamsters? I'm torn between two scenarios. In one, parents who have used the contraception have to buy their lonely only child a hamster for company. In the other, parents use pets to distract their children and avoid interruption while they take advantage of the wider availability of contraception.

I wouldn't pretend this kind of frivolous trawling tells me anything at all about what people were thinking when they wrote the books. But Steven Skiena, in his pre-Yahoo! research, had something far more ambitious in mind than merely counting how often words appear.

'Something that we worked on was: Could we analyse all of this data and figure out how words change their meanings?

For example, the word gay doesn't mean what it meant when I was a kid. Can you plot how its meaning has changed over time? When does its meaning change? How does its meaning change?'

Skiena tells me how to approach the problem. 'There's a technique you probably heard about: Deep Learning, which we use in work going on here in the lab, on something called word embeddings, a way to analyse large amounts of text and figure out a model of what each word means, by viewing it as a point in space.'

What Skiena is talking about is an algorithm that uses mathematics to model how words relate to one another in use. Which is more intuitive than it might initially appear. If you're deciding which word to use inside a greetings card, you might think: The card has HAPPY on the front, so I want another word that's close to happy in meaning, such as merry or glad or joyful.

'So we like to analyse a lot of text, and then for each word put that point in space such that words that were similar were close to each other in space. You can imagine if they were stars in the universe, it'd be a constellation of colours, a constellation of city names, a constellation of verbs … You can do this by word-embedding techniques, learning techniques.'

Of course, the machine isn't learning the language in the way that we learn a language, by using it in the real world. Skiena trains it first on five-word phrases.

'You try to train a system that is good at distinguishing a real phrase of five words from one where they change one of the words to a random word', he explains.

'Suppose "the cat in the hat" is five words and now I change the word hat to a random word. "The cat in the … yellow" is not nearly as likely to be a real phrase as "the cat in the hat". You can imagine trying to train a computer that is good at distinguishing between the real phrases and corrupted phrases.'

The computer goes through millions of five-word phrases, sorting them into real and corrupted, getting feedback about which are right and wrong.

'In the course of doing this, it's moving these points, these stars, that represent what the words are, it's moving them around. So it gets better and better at telling the difference between the real phrase and the bad phrase', Skiena says. 'Along the way the stars are going to figure out something about grammar, about what words occupy similar spaces, and this is a way to get some understanding, in some sense, of what words mean.'

What words mean is a big question. And I am sceptical that a computer churning through five-word phrases, even millions of five-word phrases, can truly understand human language. The exact workings of the computer's reasoning process are a mystery, because it teaches itself as it goes along. It's a black box, in the sense that stuff goes in and other stuff comes out, but nobody, not even the person who designed and programmed it, knows what happens in between.

But I do think about words, language, even ideas, through spatial metaphors. That's why I wave my hands around so much when I'm debating abstract ideas. So I feel a certain affinity with Skiena's program, even if it is working in 200-dimensional space, which I can't possibly visualise.

'The program is going to figure out where to put each word, and to a first approximation that 200-dimensional point is what the word means. Other words that are close to it in this space are presumably similar words, and do similar things. It captures a lot of what you would call meaning, maybe not everything.'

But there is one stumbling block for sentiment analysis. Americans have a reputation, which I have found mostly unjustified, for being weak on irony. I suspect we Brits flatter ourselves that we're better at irony, when most Americans do understand it, but are simply too polite, or not cynical enough, to sink to sarcasm.*

---

* If you're thinking, 'Yeah, right!' I bet you're British.

Computers, however, really don't get it. So how does Skiena deal with it, when he's trying to find out what people are thinking and feeling?

'Irony is one of these things that it's hard for a computer to get. But one of the things about big data is of course that you can be wrong a lot. If it came down to reading one tweet and the guy said –', Skiena puts on a sarcastic tone that would make an Englishman proud – '"Yeah I really think this is great!" then you're in trouble. But if you got 1 million tweets about Volkswagen and most people are not being ironic …'

Then you're going to get the right overall result, says Skiena.

'You'll be needing these…'

Companies use this kind of sentiment analysis as market research, to get a general sense of how happy their customers are, or how well their new product is going down. But the personal butler will need to know what you're thinking, what you like, and what you're interested in, not just to give you the best answers when you ask for the nearest coffee bar, but to anticipate your needs before you have them.

'Coffee is at the corner of 43rd and Broadway, and while you're there, you might want to pick up a book about Hawaii because you're going there next week,' is Ken's example.

This is where I feel completely divided. As somebody profoundly disorganised, I have long fantasised about being able to afford a PA who could not only do my admin, but tell me what to do.

This fantasy assistant would remind me that I have an event next week, and no free time after this afternoon, so I'd better do my preparation now. It would not only book my flights, reserving a window seat, but look up the train times to get me to the airport. It would know that I'm bad with names, so when I re-meet people, even ones I've known for years, it would discreetly remind me that they're called Dave.* It would,

---

* But only if they are called Dave. Otherwise that would just be mischievous.

if not pick up my shoes from the cobblers, at least remind me, just before I pass the cobblers, to collect them myself.*

This is exactly what the virtual butler promises. Siri, Cortana or Google Now can do a lot of this already. Have I embraced them? No! I reluctantly share my work schedule with a few trusted colleagues, but the idea of some algorithm knowing everything about my life, some algorithm that is no doubt aggregating all my data with millions of other users' data, creeps me out. And when I admit this, Ken says I'm not alone.

'Helpful versus creepy is a problem we all work on and worry about a lot. We've got to get it right. In fact, we don't release some of these features as soon as perhaps we'd like to because of that. We'll try it out on ourselves, we do a lot of experimenting on ourselves. We're our own lab rats here.'

It's not a problem unique to the virtual butler or the digital PA of course. Most people who have used internet searches, or social media sites, have experiences of targeted adverts that are either hilariously wrong or creepily right.

Steven Skiena has a horrible example. 'When my mother was ill and I had to start making the trip down to Florida, I started getting ads for funeral homes in Delray Beach Florida.' Less a butler, more a digital Angel of Death. Luckily, he says, 'she hasn't needed it, and she's better, and I'm very happy about this.'

This is exactly what a human assistant would not do. They might make contingency plans for rescheduling future appointments, but they would never be so tactless as to offer you funeral suggestions while your relative was sick. The perfect assistant would also have a sense of how much you want to share about your personal life, and what you'd consider

---

* I do realise that I should be able to do all this myself, and that it's pathetic to be such a child. I'm just being frank about my flaws, OK? To be honest, I'd hoped for a more sympathetic response from you.

intrusive. That's a big leap from today's question-answering applications, but it's the hope for Skiena's big data and machine-learning techniques: to move from a device that can understand you want to fly to Florida, and offer you flights to Florida, and stop when you've booked those flights, to one that can put that information in context, as a human would.

Skiena brings it back to the question of 'shallow understanding' and 'deeper understanding'.

'For example I was telling you that I could tell you something about what words mean. Happy means the same as glad. "What does it mean?" becomes a more complicated thing as I tie more words together,' he says.

'It's one thing to say I know what the word "happy" means, it's another thing to say I know what this Shakespeare play means.'

Subtle, false and treacherous

"Now is the winter of our discontent made glorious summer by this sun of York!" An algorithm practising sentiment analysis on Richard, Duke of Gloucester, at the start of Richard III would no doubt classify him as feeling happy.

By the end of the speech, he has told us in so many words that all this love and peace just makes him angry, and so he's going to stir up trouble. So our reading of his mood might change as we learn more. Or, knowing about the play, or about the history on which it's based, we might from the start feel he's speaking ironically.

And this is just the opening couplet. No two human minds would agree on exactly what the play *means*. I asked one friend, a Cambridge English graduate, and he replied with more questions. 'Does Richard choose to be evil? If he does, is he less plausible? Or more tragic, because so fatally wounded morally, as well as physically?'

Questions lead to more questions, and not one with a binary, yes/no answer:

'Maybe we say he doesn't quite choose freely, even though he says he does. If he has reasons, which by the way he does explain (even though they are not all justified and he sort of knows they're not) do they justify him for us, or only for himself, or are they the lie he tells himself to fulfil a deeper evil that he doesn't acknowledge properly (because people don't)?

Thus, ultimately, what *is* evil?'

There are so many layers of meaning in a play, the symbolism, the music of the language, the resonance we get from knowing the modern usage of the words and learning something of their archaeology by hearing how Shakespeare uses them, the narrative of events, the emotional trajectories of the characters ... and all this without even starting on performance, staging or design.

No digital assistant can convey the experience of watching a play.

Steven Skiena helps explain why.

'Historically, these big data systems, these machine learning systems, usually are of the form: examples with answers. You've asked people, so you know what the right answer is, and you're good at predicting the right answer. Problems that don't have right answers are not so easy to train systems for.'

And so many problems in the real world don't have a right answer.

'My college students are still a lot smarter than any intelligent systems,' says Skiena. 'There are certain tasks that happen to be relatively easy to build systems for, and there are other tasks that don't necessarily look harder, but are much more open-ended.'

'For example, playing chess at a grandmaster level. A hard thing for me or you, but it is relatively easy to build a computer system that would beat us all in chess.'

But, says Skiena, 'When you are talking about the example of irony, that is a much harder task for a computer than the problem of playing grandmaster chess. Part of it is a question of how big the universe of humanity may be, that you need to have associated with the task. Chess is a very well-defined game. It doesn't exist outside of its world.'

For irony, however, 'You have to model what Kim Kardashian is, understand certain snarky references and things like that, so what seems like a simple problem involves much more world knowledge.'

On my way out, Steven Skiena shows me his desk, from which he can see the ball in Times Square that drops at midnight on New Year's Eve. It's reflected in another glass skyscraper, but it's still quite a view. I can tell he plans to be here to see it. Working, no doubt, but with an eye out of the window on the bigger universe of humanity.

CHAPTER FIVE

Big science

There are 89 cubic kilometres of water in Lake Geneva. That's 89 trillion litres (19.6 trillion US gallons) of water. I can barely hear the traffic above the roar of lake water through the sluices. As I look across the grey expanse, two thoughts occur to me. The first is that carrying my 1-litre ($1^{3/4}$-pint) bottle of drinking water feels a bit silly now. The other is that, if it keeps raining like this, Lake Geneva will soon contain 90 trillion litres of water.

I've walked down to have a look at the lake, and the famous Jet d'Eau fountain, on my way to CERN, which is a tram-ride away from Geneva station. Well, the main buildings are a tram-ride away. I suspect that at least part of the experimental equipment is somewhere beneath my feet.

There's 27km, nearly 17 miles, of tunnel in a big loop under the Swiss/French border, where the subatomic particles go flying round at mind-boggling speeds[*] before colliding: the Large Hadron Collider, LHC. The hadrons are tiny subatomic particles, it's the whole engineering set-up that is the 'large' part. And the amount of data that it chucks out every time they fire it up is big by anyone's standards.

When they cross the beams,[†] the detectors generate 40 million megabytes of data per second, which is more than they could store, let alone analyse. But, as with so many

---

[*] A proton at full speed completes 11,245 circuits of the 27km-long (17 mile-long) tunnel every second.

[†] 'Cross the beams' is my phrase. A real CERN scientist is more likely to say 'run the experiment' or 'induce proton packet collisions' or something more technical. Though crossing the beams is exactly what they do.

things, it's not the size of your data that counts, it's what you can do with it.

Think of my pathetic bottle of water. Rain is dripping off it now, and soaking through my coat. But I'm still holding it, because it's the volume of water that I can use for my specific purpose. It's been selected, processed and packaged to keep it free from contamination, so I can drink it. That's what makes it more useful to me than all those billions of litres rushing past my feet or plopping off my nose.

If I tipped this litre of drinkable water into Lake Geneva, it wouldn't take long for the currents to mix it with the other 89 trillion litres. Some of it might be squirted through the Jet d'Eau, which puts 500 litres (110 US gallons) per second into the air. It would certainly be diluted with the rainwater.

Imagine what a task it would be to go through all the water in the lake and reassemble the original water molecules back into this bottle. Like one of those fairy tale challenges where the heroine has to separate a mountain of peas and lentils, one by one. And I think: That's about the scale of the task they've set themselves just up the road, at CERN. Only it's not water they're sorting through, or lentils, it's data.

Asking the big (and very tiny) questions

I was expecting something a bit more like a mine. I have my helmet, my pass, and my authorised Underground Tour leader, but when the lift arrives, Piotr gets in and presses the Level -2 button. 'It's just an elevator,' he says as we descend 100m (328ft) to the tunnel.

Piotr Traczyk works on CMS,* one of the two main experiments here at the LHC. Like its sister, or rival, ATLAS,

---

* Compact Muon Solenoid.

The *solenoid* is a coil of wire that produces a magnetic field when electricity flows through it. In this case, it has a 7m (23ft) diameter and produces a magnetic field about 100,000 times stronger than that of the Earth.

CMS is essentially a system of detectors that record what happens when the beams of protons collide. By measuring what is thrown off, at what speed and what angle, the scientists can reconstruct events that only lasted for fractions of a second. The brief appearance of a Higgs boson, for example.

The first thing we reach when we leave the lift is a roomful of computers, rack after rack of hardware with blue cables and winking lights in all directions. It looks like a scene, either from a 1960s documentary about the new, compact computers that only take up one room, or from a James Bond villain's underground HQ. Piotr has to raise his voice above the cooling fans. 'This is where the Level 1 Trigger System lives. The trigger is the system that reduces the amount of data to manageable size, by discarding the collisions that we believe, or actually, the Trigger believes, to be uninteresting.'

If the word 'trigger' makes you think of a cull, you're right. The equipment in this room is programmed to identify potentially interesting results, and throw the rest straight into the virtual bin. This has to happen in millionths of a second, far faster than a regular PC could make those decisions, let alone a human mind. And this is just the first stage. 'There's another room directly above this one,' says Piotr.

(Cont.) A *muon* is a type of subatomic particle, produced when the proton beams collide, which may be an indication that something important has happened. For example, the elusive Higgs boson decays into four muons, so seeing four muons racing away from the scene of the crime is a strong suggestion that a Higgs boson was briefly there.

The CMS equipment is 15m (49ft) in diameter, 21m (69ft) long, and weighs 12,500 tonnes. So I have no idea in what sense it is *compact*. Though it is smaller than ATLAS, the experiment on the opposite side of the ring of tunnel.

'The first level has to run very fast. The blue cables are fibre optics: the other end is physically on the detector. The data from the detector comes directly into these racks. One collision is one megabyte, so 40 million megabytes per second enter this room, 100,000 megabytes per second leave this room.'

Piotr displays his impressive recall of many large numbers. 'Then, the 100,000 events per second get sent upstairs to the surface, and there we have a farm of 2,000 normal PCs. What we do up there is reduce the 100,000 per second to 100 per second.'

If you're losing track of all the zeroes here, imagine taking half the water in Lake Geneva and reducing it to only 40 Olympic swimming pools. Which still seems like a lot of water, so that might not help. OK, think of taking two-thirds of the population of the UK, and reducing it to 100 people, within a second. Which is pretty much what happens to TV sports viewing figures every time a British team gets knocked out of a major sporting tournament.

At this point, as I'm trying to absorb the scale and speed of all this data processing, my cellphone rings. Considering how often I can't get a signal in London, where I live, I'm surprised to get one this far under France. Piotr tells me that the whole tunnel has both Wi-Fi and cellphone coverage, for convenience and safety.

For example, each of these blue cables had to be connected by a human being, and at one point he was in this room with his spanner, while his colleague attached the other ends of the cables to the detector, communicating by cellphone. And yes, he says they do sometimes find two cables swapped over. It's nice to be reminded that all this advanced science and technology still relies on human beings tightening things up while shouting, 'It's the blue one!' 'That doesn't help, Piotr – they're *all* blue!' into their cellphones.

The next set of doors is marked with radiation signals. Piotr seems unworried by this. He also tells me the helmet is mainly to protect me from hitting my head on one of the pipes running along the ceiling. He's significantly taller than I am. But before we get to the main tunnel he has to get us through an eye-scanner security gate. Maybe it's more X-Men than James Bond.

'Beyond this door lies the CMS experiment,' says Piotr, and with a little theatrical flourish he lets me into the main cavern.

It's massive.

I'm visiting while the LHC is closed for an upgrade, which means the CMS is opened up like two halves of a bagel, with a hole through the middle where the beam pipe would normally carry the colliding particles. So I can see the five-storey-high ring of electronics: red, silver and shiny copper detectors linked by a dense net of multicoloured cables and copper gas pipes. It's like a twenty-first-century rose window, in an underground cathedral to physics, all brightly lit and framed by green metal walkways at all levels.

Piotr tells me that ATLAS is even bigger, but CMS is heavier. 'And prettier,' he adds. It is very pretty.

Each of the 11 slices that make up CMS has different detectors, to measure the track of different particles thrown off by the collisions. Part of the challenge is to reassemble these traces to create a reconstruction of each individual collision. It's like Crimewatch for particle physics, or some kind of accident investigation: 'The first proton came around this corner at one minute after midnight, and collided with the second proton, which was travelling at 1,093 million km per hour (679 million mph), well over the speed limit.'

And, to make that task harder, this requires a very precise measurement of the position of each detector when it recorded the data. 'You might think that, as we are in Switzerland, this is no big deal.' says Piotr. But all these precision instruments are set in thousands of tonnes of iron, around a magnet 100,000 times stronger than the Earth's magnetic field. Things tend to move around.

So as well as detectors, the CMS needs position detectors to measure the positions of the detectors, using lasers and mirrors. And those position detectors are throwing out yet more data that has to be collected and analysed.

So much data comes out of CERN, even after the ruthless culling of the Trigger process, that the work of processing needs to be shared among 170 computing centres in 40 countries. The Worldwide LHC Computing Grid is the biggest computing grid in the world, handling 30 million gigabytes per day. If you're thinking about how slowly your home broadband downloads cat videos, don't worry: CERN has its own private network of optical fibres handling 10 gigabits per second.

Over 10,000 physicists from Korea to Illinois, Moscow to Barcelona, have access to that data almost as fast as it pours down the cables from Geneva. The blue optical cables that Piotr and his colleagues attached to the detectors are only the start of a long journey. If their data really were the water in Lake Geneva, it would be pouring along massive pipes at near light speed, and scientists thousands of miles away could be pouring it into a glass within minutes.

Except the lake wouldn't be getting any emptier. It would be filling up with new data faster than the real lake was with this morning's rain.

The sheer scale of the data CERN produces is crucial. The Higgs boson, for example, is a rare and elusive subatomic thing, whose unlikelihood means you need billions and billions of bits of information from millions of experiments to be sure you've glimpsed one, let alone measure some of its characteristics. Equally important are the speed of processing, faster than an unaided human could do it, and the fact that thousands of scientists can collaborate on making sense of the shared pool of data.

Without big data, CERN would never have found evidence that the elusive Higgs boson particle exists. It's an obscure corner of physics, but not finding it could have meant having to rethink the fundamental building blocks of our universe, and rewrite the physics textbooks. So physics teachers all over the world, though they might struggle to explain exactly what a Higgs boson is, are grateful that they won't have to go back and learn an entirely new model of reality before the start of next term.

But the physics teachers shouldn't relax too much. Over in the astronomy corner, big data is being used to investigate some equally fundamental questions. How did the universe form just after the Big Bang? What is dark energy? What are the limits of Einstein's general theory of relativity?

Astronomy has always been a very mathematical science. Remember Laplace developing statistics to help model the underlying laws of motion from his observations of the stars and planets? Today, it's moved so far from just looking through telescopes at visible objects that many astronomers can hardly ever find an excuse to go to observatories in Hawaii, South America and the Canaries, or even to stay up all night gazing up at the stars.

They're more likely to find themselves writing computer programs to analyse digital data coming in from thousands of miles away, or even from instruments that have been sent out into space. Today's astronomers are often distinguished not just by the objects they study, like solar astronomers who study the sun, but by the type of data they analyse.

Radio astronomers collect and study radio waves from the universe. They use large dishes like the ones at Jodrell Bank in north-west England, similar to your neighbour's satellite dish that picks up Italian football games and those foreign movies with all the sighing and groaning. But what the astronomers receive is more like the background noise between stations, which takes a lot of maths to decipher.

No wonder they have to pass the time by thinking up imaginative names for their equipment like the VLT (Very Large Telescope) the ELT (Extremely Large Telescope) or the OWL (OverWhelmingly Large) Telescope. I'm not making these up. I'm a little disappointed, in fact, that the next massive international astronomy project won't be called the WYLATSOTT (Will You Look At The Size Of That Telescope) but the SKA – the Square Kilometre Array.

In radio astronomy too, size really does matter. Because radio wavelengths are longer than visible light, it takes a much bigger dish to collect enough to make any kind of picture. But with a big enough dish, radio astronomers can find some quite exciting things. Jodrell Bank's Lovell Telescope, 76m (250ft) across, enabled Jocelyn Bell Burnell to discover the existence of pulsars, fast-spinning neutron stars emitting a regular pulse of radio waves. In the 1950s and 1960s, Jodrell Bank tracked Sputnik and other early space missions.

The largest radio telescope dish on the planet, at Arecibo on the Caribbean island of Puerto Rico, is 305m (1,000ft) across, and is built into a natural hollow in a mountain. Its size makes it so sensitive that it can detect signals that have been travelling across the universe for 100 million years. If that doesn't impress you, it also featured in the James Bond film, *Goldeneye*.

The SKA won't literally take up $1km^2$ (0.39 $mile^2$). Instead, an international consortium of scientists plan to use the same kind of fast, accurate, data-heavy computing that CERN gets from its grid.

Instead of a single, WYLATSOTT receiving dish, SKA will consist of an array of smaller dishes and receivers spread across two sites in Australia and Africa. These detectors will generate more than 10 times the amount of data carried by the whole internet today. Like CERN, it will combine the data received from the different detectors to build up a picture.

To put together signals received thousands of miles apart, it will need to make allowances for differences in time as well as space.

This will require a computer three times as powerful as the most powerful supercomputer in existence in 2013, and thousands of miles of fibre optic cable. And, like CERN, SKA will need to cull the data before sending it out to teams of scientists all over the world.

Like most large scientific projects, SKA will take many years to complete, and probably won't be fully operational till around 2025. That should give them plenty of time for supercomputers to become three times as powerful as they are today. If you're currently at school, you could get a job analysing some of the first data the SKA spits out, or writing the computer code that will do it for you.

In some ways, physics and astronomy are the easiest targets for big data. I'm not saying that studying elusive particles that exist for fractions of a second, if at all, or finding dark matter whose main characteristic is that it can't be seen or heard, is easy. But the problems they throw up are the types of problem that big data is best equipped to tackle.

Subatomic particles and galaxies may be very small, or very far way, but they tend to behave according to the laws of physics. These laws, being mathematical, talk the same language as computers, at least as closely as Brits and Americans talk the same language.

As soon as living things are involved, things get more complicated, as we found with Professor Eamonn Keogh's insects in Chapter 1. And yet, as we also found, big data can help spot patterns in living things as well as inanimate lumps of matter.

Big brains

After he'd done his PhD in neuroscience, John D. Van Horn became a researcher. One of his first jobs was to go out and buy the biggest hard disc his lab could afford, to store all the data their research was going to generate. 'People would come to our lab to look at the hard disk in awe,' he says, 'because it was four gigabytes! Wow! Amazing! At that time, I thought four gigabytes was infinity.'

I don't like to ask how long ago that was, because the smallest iPod Nano had 4GB of storage back in 2007. My cellphone has twice that much memory. And today, Professor Van Horn's neuroscience lab at the University of Southern California stores, 'many, many petabytes of data'.

There's a two-way relationship between how much detail the technology can deliver and how complex the questions are that the researchers want to ask. Van Horn started out measuring blood flow in the brain, using equipment that could give one image of the whole brain every eight minutes. Now, he can use functional MRI to give him three or four images per second, not quite real time compared to brain activity, but fast enough to see changes as the person in the scanner thinks, or moves, or reacts to what they're seeing or hearing.

The human brain is notoriously complex and variable. Every brain is unique. Put two different people in a brain scanner and their brains will be different. Recognisably the same organ, but unique in the detail of its folds and wrinkles. Ask them to do the same task, and you won't see exactly the same pattern of activity in those two brains. Similar, probably, with the visual cortex at the back of the skull engaged in making sense of what they see, or the language areas on the left hemisphere analysing and producing words, but human beings have an impressive ability to find different ways to get the same results, even inside our own skulls.

This is where the pattern-spotting ability of AI can be useful. Show enough brain scans to an AI algorithm and it can learn to sort them according to whatever criteria you gave it in the training set: male or female, musician or non-musician, normal or psychopath.[*]

Because it has taught itself the best ways to distinguish between categories, we don't necessarily know what the differences are. But whatever its criteria, a computer can classify human brains by looking at scans. This could have all sorts of uses.

Professor Van Horn, for example, is looking for rare genetic variants associated with brain diseases, such as autism and Parkinson's. He does this by linking a genetic database to his database of brain images.

At Imperial College in London, Professor Paul Matthews also uses big data methods. But, for a Head of Brain Sciences, he's surprisingly equivocal about the brain scans in his database.

'Yes, I'm certainly interested in the brain data,' he says, 'but it's only one piece in a jigsaw. The public perception of it, and even some doctors' perception, is that it is somehow special. But it's not really. It's a piece of data. It's a particularly rich piece of data, but it's just a piece.'

For Professor Matthews, it's the ability to link the scans with other sources of information that marks the move, 'from large data to big data'.

'Large data is data in which we have collected lots of a single type, or a limited type, of information, and we are using it within the context that was originally envisioned. Rather than studying the size of the brain in 20 subjects, we're going to study it in 1,000. So we're scaling up a simple idea.'

---

[*] We can't say that because these category differences are physically recognisable in the brain, they must be innate. Our brains change, physically, in response to how we use them, so becoming a musician, or a man, or a psychopath, might have involved repeated actions that reshaped the brain.

Going beyond sheer scale, says Matthews, 'Big data is where we're not just collecting one or a few types of data, like brain scan, age and gender: we're collecting all the things we can possibly think about and more, and we're looking to relationships between them that we might not have anticipated. Moreover, we're starting to link between different datasets.'

This is the kind of thing Van Horn is doing with his genetic database. But Paul Matthews looks far beyond medical records.

For example, 'because you have their postal code, you can go back to meteorological records and say something about exposure to sun, or to particulates in the air, that these individuals had. So using bits of data from one dataset to link into other datasets, suddenly the picture you have of a person's life is expanding rapidly.'

Does this ability to bring together many different dimensions of a person's life, and set a machine to plough through in search of new patterns, mean the researcher's role is fundamentally changing? Some have claimed that big data represents the end of theory, as computers will produce answers before we even know which question to ask.

Remember your school science lessons, where you had to write down a hypothesis, then test it with the apparatus, note your results, and state your conclusions?* Is that really an outdated approach? I put it to Professor Matthews.

'Is it all about trying to find relationships blindly or is it about hypotheses? You can do either,' he tells me. 'That's the beauty of these datasets.

I might hypothesise that people on statins are less likely to show progression in multiple sclerosis, which recent trials suggest. I can go into the dataset of people with MS and

---

* Mine always started with, 'allowing for experimental error ...' as the results were never the ones we should have got. That's one reason I became a writer instead of a scientist.

identify those who have used statins and those who have not, balance them for other factors, and ask the question, which population did better? That would be a hypothesis-led association study. The alternative is to use sophisticated data-mining approaches to allow clusters of features to be identified. It's always back and forth.'

So scientists, relax, you're not out of a job yet. Think of your big data AI as a lab assistant, that might sometimes notice interesting new results and point them out for further research, but the rest of the time is doing your donkey work.

Genetic codebreakers

In genetics, like neuroscience, the mass of data goes well beyond what the human computers of a century ago, with their pens, paper and mechanical adding machines, could have handled.

The human genome is made up of around 20,000 genes, stored on 46 strands of DNA* called chromosomes, found in the nucleus of almost every cell in the body. The fact that different types of cell – skin, bone and brain – contain the same genetic blueprint should be a clue that the whole process of making a human being is not as simple as following a recipe book.

What a recipe book it would be: a book with 46 chapters and around 6 billion letters. Though chromosomes come in pairs, so it would really be a book with 23 chapters, each in two alternative versions. Except if you're cooking a boy, in which case one of the chapters would have an X version and a shorter Y version that is completely different and mainly

---

* Deoxyribonucleic acid, a chemical that comes in the form of two strands twisted together, the famous double helix that has become the universal symbol for genetics.

about how to make testicles and sperm. It would be a book unique to the person being made, unless they're an identical twin,[*] to which each parent has contributed 23 chapters. Typos may have crept in. Sometimes the two versions are incompatible. Perhaps Dad uses pints and pounds, while Mum uses litres and kilogrammes.

Somehow, our imaginary cook has to decide which word to take from Dad's chapter, and which from Mum's, trying to avoid any misprints that could make the whole dish inedible or even poisonous. And there isn't even a cook.

There's still much we don't know about the cooking process, but scientists published the whole recipe book in 2003. Two rival teams, in the UK and US, competed to be the first to map the human genome at the end of the twentieth century, but the final version was a collaborative effort that also included scientists in Germany, Japan, France and China. It cost several billion dollars and took over 10 years. It's now freely available to anyone, though it's not much of a read if you're not a geneticist, being made up entirely of the letters A, C, G and T.

And, as you may have guessed, it's not one genome that describes every single person on the planet. It's a representative version that combines material from a number of volunteers, much as editions of the Bible, or the works of Shakespeare, draw on a number of sources to produce what looks like an authoritative edition.[†]

---

[*] Anybody who works in genetics hates the term 'identical twin' because they're not really identical. They prefer 'monozygotic twin'. The word monozygotic means that both twins came from one fertilised egg, or zygote, and has a maximum Scrabble score of 261.

[†] Though Craig Venter, who led one of the original research teams, has since sequenced his own genome in its entirety, and transmitted it into space. So one of the first things the SKA detects could be an army of cloned Craig Venters, invading from a more technologically advanced galaxy that has decoded the recipe book for Craig Venter and worked out how to grow human beings in a factory.

Rather like big data, the mapping of the whole human genome was hailed as the answer to all our problems: We would understand all diseases and be able to tailor cures to each individual. Decoding a person's genetic blueprint would be like looking into their future. Science would have godlike knowledge of all things. And of course, it turned out to be a lot more complicated than that. We can tinker with the DNA of simple life forms, such as yeast or bacteria, quite well. Humans are a bit more tricky.

Nevertheless, by identifying specific genes linked to specific disorders or diseases, we have got closer to preventing and treating them. People whose families have a history of illnesses such as sickle cell anaemia, who are wondering if there's a danger they'll pass on the illness to any future children, can take a test and find out. Parents may have the option to bypass natural conception with *in vitro* fertilization (IVF), what's sometimes called a test tube baby. This gives doctors the possibility of screening embryos for certain faulty genes with a technique called pre-implantation genetic diagnosis (PGD). Then parents can choose to implant an embryo that is free from a life-limiting or fatal disease.

Given the unfortunate history of eugenics, which brought together the new fields of genetics and statistics only to find itself in bed with genocide, you may feel wary about this. But there's no inevitable path from the human genome to genocide. They share a linguistic root, but so do generator and genitals, and we don't want to stop using either of those.

Medically, the science of genetics has immense promise. Thousands of medical conditions are caused by a fault in a single gene, in the way a broken arm can be caused by a single fall from a horse. Cystic fibrosis, Huntington's disease and sickle cell disease are among them. Having one of these conditions doesn't make anybody less of a person, but I'm glad I was lucky enough not to draw any of those tickets in the genetic lottery, and I'll be even more glad if medical science can help future individuals be equally lucky.

Gene therapy, though still in early stages, offers the promise of fixing a faulty gene in a living person.

But these relatively simple genetic faults are just the beginning. More often, many different genes work together to produce different effects or predispositions. This is where harnessing genetics for medical benefit starts to get tricky, and where big data can start to be properly useful.

In many cases, having this or that genetic mutation doesn't mean a disease is certain, only that levels of risk have changed. Actress Angelina Jolie chose to have a double mastectomy, the surgical removal of both breasts, though there was no obvious sign she had breast cancer. Knowing that she had a BRCA1 genetic mutation, which brings an increased chance of developing the disease, she had decided to take preventative action.

Jolie herself reported that doctors told her she faced an 87 per cent risk of developing breast cancer, though on average those with a BRCA1 mutation face a 65 per cent lifetime risk. Scientists may quote a range of risks from 45 per cent to 90 per cent associated with a faulty BRCA1 gene. Even 45 per cent is significantly higher than the 12 per cent overall risk that a woman picked at random from the general population will develop breast cancer at some point in her life.

Women with a family history of breast cancer, like Jolie, whose own mother died of the disease aged just 56, can take a genetic test that looks specifically at genes like BRCA1 and BRCA2. Knowing whether they have a version of the gene that increases their own risk gives women choices about reducing that risk through surgery or in other ways.

This kind of knowledge is possible because an individual's genes can be mapped far faster, and more cheaply, than when the first human genome was published a decade ago. In 2004, sequencing a new human genome would have cost around $20 million. Ten years later, that cost had fallen below $5,000, and now a company called Illumina can do it in a few days for under $1,000.

This is not just because all those scientists who worked for 13 years on the very first human genome are getting quicker with their pipettes and microscopes. It's thanks to improvements in hardware, software and information handling. Biology is moving rapidly from being all about the specific, fiddly, squishy thing you have on your lab bench, or your microscope slide, to being all about masses of data. And this opens up new possibilities.

Nobody could have looked at Angelina Jolie, seen into her future, and known for sure that she would or would not develop breast cancer. And if she did develop it, nobody would have been able to tell that the mutation in her BRCA1 gene was definitely to blame. But by noting a correlation between having that mutation and going on to develop breast cancer in thousands of other women, scientists were able to put a figure on her increased risk. And, knowing that figure, she was able to make her own decision about whether to take preventative action.

Most breast cancers are not linked to faults in this particular gene. BRCA1 is implicated in less than 10 per cent of breast and ovarian cancers. If you do have a faulty version of a BRCA gene, your chances of going on to develop one of those diseases is higher, but it's not certain. You might be lucky.

Is it more than luck? Is there more we can do to avoid illnesses like cancer? That's the kind of question that big data can help us address. Illumina, the company that can provide the $1,000 human genome, is contributing to a UK study called the 100,000 Genomes Project.

It will target cancers and rare diseases, sequencing the genomes of patients, their families and, in some cases, of the cancerous cells.

By linking the genetic information with what's known about the medical history of the individual, scientists hope not only for a better understanding of how genes affect our health, but also of how to treat illnesses more effectively. For example, by knowing the genetic profile of an individual's cancer, doctors can know which chemotherapy is likely to be most effective.

But just as reading a recipe book is no guarantee that your dinner will taste perfect every time, reading a human genome won't tell you everything that will happen to that person. Some diseases are affected by environmental factors, and others are down to sheer bad luck.

Big danger?

When doctors, or medical researchers, talk about environmental factors, they don't just mean whether you're breathing clean air or getting sunburn because you grew up next to an Australian beach. Any factor that's not in your genes when you're born may be called environmental.

So smoking and living in Cornwall would both be described as environmental factors, even though deciding to smoke was a choice you made, unlike being born in Cornwall.

In Cornwall, the levels of background radiation are much higher than in most of the UK, simply because of the type of rock that makes up that part of south-west England. Radon, a radioactive gas emitted naturally by the uranium in rocks, causes over 1,000 deaths from lung cancer every year in the UK. That's a higher death toll than drink-driving. That said, even quite high doses of radon gas don't increase your risk of lung cancer anywhere near as much as smoking cigarettes.

In practice, it is hard to make a clear distinction between environmental factors and things that you have chosen to do. You could choose to move away from Cornwall when you reach the age of 18, for example, instantly reducing the dose of radiation that hits you every day. But by moving to London to avoid death by radon gas, you could expose yourself to other environmental risks, from air pollution to loneliness.

You could stay in Cornwall and fit radon protection to your house, as it's gas accumulating indoors that carries most of the increased risk. You could choose to reduce your exposure even more by working outdoors, though farming and fishing are two of the most dangerous professions. Around one in 10,000 Britons working in agriculture, forestry and fishing will die on the job in any given year, making it the riskiest sector to work in.

So, even knowing the data on risks to your health, it's not easy to apply that knowledge to making decisions as an individual. However, if you are a public servant looking to reduce the numbers of deaths across the population, it can be helpful to know where to apply your efforts: offering radon gas protection to homes in the highest radiation areas, for example, or trying to cut down the number of people who smoke.

Smoking is one of the big success stories of modern public health. Or rather, stopping smoking is. Today it seems amazing that millions of people smoked cigarettes without any sense that they might be harming their health. But in the mid twentieth century, when doctors and scientists looked at the rapid rise in cases of the previously rare disease, lung cancer – a 15-fold increase in 25 years – smoking was barely in the list of suspects.

If this seems bizarre, remember that the 1950s were a very different age. Turing's imaginary robot child escaped the everyday household chore of filling the coal scuttle only by virtue of lacking legs. Most homes were heated by burning coal, wood or peat indoors. City air was darkened by smoke from all those chimneys, as well as from industry. Until the 1980s, when a big cleaning programme began, buildings in Britain's cities were black.

The great smog of December 1952, a blend of smoke and fog with high levels of sulphur dioxide, caused an estimated 4,000 premature deaths in British cities. The Clean Air Act of 1956 authorised smoke-free zones in London and northern industrial cities, with controls on what industries could emit, and a shift for households from coal to smokeless fuel. The US passed its own Clean Air Act in 1963. And the gradual shift from domestic fires and stoves to gas or electric central heating must have made a difference over the next 40 years or so.

So the visibly polluted air that city-dwellers breathed in the 1950s was an obvious suspect for causing lung cancer. Other potential villains included the rising number of petrol-driven vehicles, dust given off by the new asphalt-coated roads, or even residual effects from having been gassed in the First World War, as men were more likely to develop the disease.

Some people suggested that improved diagnosis and increasing lifespans were at least part of the explanation: people were living longer, and developing the slow-growing disease instead of dying earlier from something else, or deaths previously thought to be caused by tuberculosis or other lung diseases were now being identified as cancer. A study in Germany in 1939 suggested that lung cancer patients included a smaller proportion of non-smokers than the general population did, but research done in Nazi Germany was not the first place most researchers wanted to look.

Today, this would be seen as a classic big data problem. Over a population of millions, we can identify from health records which people develop lung cancer. All we have to do is find the risk factor with the closest link to the disease. But which other information would you include in your search?

With hindsight, armed with your twenty-first-century knowledge that smoking has been tried and convicted, you might suggest medical notes: surely doctors asked whether patients smoked? Perhaps not, 80 per cent of middle-aged men in the UK smoked in the 1950s, including most of your patients who did not have lung cancer, so only one in five of your patients was likely to answer 'No', including the majority of your patients who did not have lung cancer.

How else could you find out? Today, supermarket store cards could help, if smokers bought cigarettes with their groceries. Though you wouldn't necessarily know which member of the household was going to smoke them, or if they were bought for a housebound older relative, for example. Perhaps you could analyse the photographs people post on social media, using an algorithm to detect the presence of cigarettes. All these are now technically possible, if not always socially acceptable.

Without the hypothesis that smoking was to blame, however, you'd have to cast the net much wider.

To check for the effects of air pollution, you could look at addresses. Today, we have postcodes, which makes it much easier to put people into groups according to where they live. You could use GPS and other location-recording features on people's mobile telephones to track where they spend their time. Are they walking or cycling along polluted roads? Do they spend eight hours a day working near, or in, a chemical plant? You'd need to get their permission to access these records, of course, but many people are already using their own apps to keep track of where they go, whether they walk or cycle, and so on.

Back in 1947, however, Austin Bradford Hill and Richard Doll had to use statistical techniques to test their hypotheses.

The method they used was a *case control* study, a way to compare one group of cases, people who do have the disease, with a control group who don't have the disease. In Doll and Hill's study, they chose both groups from London hospital patients. The 649 patients with lung cancer were compared with 649 patients admitted with other illnesses.

Not all case control studies manage to recruit equal numbers of cases and controls, but statisticians can still draw useful conclusions from studies with unequal numbers, using the *odds ratio*. By splitting the two groups according to their exposure to risk factors, researchers can calculate the odds of getting the disease for exposed and non-exposed groups.

Let's see how that works with some made-up numbers. Suppose that 45 of our overall group say they've lived in Cornwall. Of our Cornish selection, 25 are in the lung-cancer (case) group and only 20 in the not-lung-cancer (control) group. So more of our imaginary Cornish patients have lung cancer than not.

Does this prove that living in Cornwall is a risk factor for lung cancer? Not necessarily. We have to compare the odds of getting lung cancer if you have been exposed to Cornwall (25:20, or 5:4) with the odds of getting lung cancer if you haven't been exposed to Cornwall.[*]

Of the 1,253 people who have not lived in Cornwall, 624 are in the Lung Cancer group and 629 in the Control group, who didn't get lung cancer. So the odds of getting the disease if you were *not* exposed to Cornwall are 624:629, roughly 35:36.

These odds seem high, don't they? Odds of 35:36 is almost evens, let alone 5:4. Those odds are almost certainly much higher than you'd estimate.

This is because our sample is not a random sample of the population. We selected 649 lung cancer patients, and 649 people without lung cancer. So we started out with a 50:50 chance that a person picked at random from our group will have lung cancer: evens, in betting terms. This is much higher than in the real world, where even a smoker has about a 15 per cent risk of getting the disease,[†] giving odds of 3:17.

---

[*] 'Exposed to Cornwall' isn't a very flattering phrase for a rather beautiful part of south-west England, but it's the kind of language medical statisticians use.

[†] This is one reason a case–control study can be better for relatively rare diseases. The alternative is a cohort study, where we take a group of people before any disease has developed and watch them move forwards through time, like a marching cohort of soldiers. We could sort them according to whether or not they've lived in Cornwall, and see how many of each group go on to develop the disease. (Cont. p.155)

In order to make a useful comparison, we need to compare the odds of being in the lung cancer group, if you come from Cornwall, with the odds of being in the lung cancer group if you don't come from Cornwall. We find the odds ratio by dividing one by the other. If the odds were the same for both Cornish and non-Cornish, the odds ratio would be one.

We divide 'odds of lung cancer, with Cornwall' by 'odds of lung cancer, without Cornwall', giving us approximately 1.26, or one and a quarter. So your odds of getting the disease if you come from Cornwall are about a quarter higher than if you don't. Must be all that radon gas.

Now, I must remind you at this point that I made up these figures. I'd hate to be responsible for a dozen deaths on the A303 road as thousands of people leave Cornwall in panic. Remember, too, that saying your odds are a quarter higher is a bit meaningless if you don't know your baseline risk of getting lung cancer.

Among 100,000 non-smokers, around 16 would die of lung cancer. Using my made-up numbers, moving them all to Cornwall would push that up to around 20 in 100,000, four extra deaths. It would also increase the population of Cornwall by almost one-fifth. And the increased risk is one in 5,000, which is still pretty long odds.

The problem is that only a small proportion of either group are likely to develop lung cancer. So we would need a very large group of people to be confident that any differences were linked to the Cornwall/not Cornwall difference, and not just natural variation. From a random sample of 1,298 people, you would expect fewer than 100 cases of lung cancer over a lifetime, so it might be hard to see a clear difference in risk linked to Cornishness, even after waiting 70 years or so, and most lung cancer is diagnosed in people over 65. If you include other factors, such as smoking, you need an even bigger cohort. Using big data techniques makes it much easier to do a cohort study, because you can find a larger sample of people whose past data was already collected, and then find out how they are doing now.

So I didn't mean to cause alarm. I did it to show you how researchers can work backwards from groups of patients to estimate the risks to whole populations of exposure to certain things. In this case, Cornwall.*

This method is what Doll and Hill used to find that smoking cigarettes raised the risk of getting lung cancer. And the odds ratio they found was much more dramatic than our 5:4 ratio from my made-up figures about Cornwall.

Most of their patients smoked. Most British men did in 1947. So their exposed group of smokers was 1,269, much bigger than the non-exposed group of just 29 non-smokers. 647 of the smokers were in the cancer group, but only two of the non-smokers.

They calculated the Odds Ratio for smoking versus not smoking, and got an answer of 14.04. Smokers were 14 times as likely as non-smokers to develop the disease. You can see why both Doll and Hill gave up smoking at once.

Their later work showed an even stronger correlation between heavy smoking and lung cancer, with odds ratios between 20 and 50, compared to people who'd never smoked. They also found the link was dose-dependent: the more you smoke, the higher the risk.

It took some time for the link between smoking and lung cancer to be widely accepted. Prominent scientists, doctors and statisticians were sceptical, suggesting alternative explanations such as genetic predisposition to both lung cancer and smoking. The tobacco industry was clearly reluctant to publicise the fact their product might shorten your life.

Eventually, animal studies and improving knowledge of basic biology pointed to chemicals in cigarette smoke that suppress the body's defences against cancer. But this was one case where correlation alone was enough to affect both

---

*I did, however, base my numbers, *very* roughly, on the estimated risks of radon gas.

public policy and individual behaviour, long before a causal mechanism was shown.

Lots of tiny dangers

Today's public health studies, using big data techniques, can find much more subtle patterns. Combining huge sets of data and clever statistical techniques, researchers can find links between behaviour, environment and health outcomes, especially in countries such as the UK that have comprehensive medical records across most of the population.

Their task is much harder than that of Doll and Hill in some ways, as they're working against a background of ever-lengthening lifespans and falling death rates from most diseases, even the big killers: cancer and cardiovascular disease.

So instead of one big effect for which to seek an explanation, they're looking for relatively small factors that could contribute to speeding up improvement in the population's health and longevity. Is sitting down the new smoking? Can taking statins reduce your chance of a heart attack? Should we do more about air pollution?

But beyond the technical difficulties, which big data promises to help overcome, there are wider problems for these public health researchers.

Having access to the medical records of millions of people, let alone linking them to other personal details such as where they've been, or their personal habits, risks intrusion on the privacy of individuals.

Finding a link does not always mean you have identified a cause: remember Farr and his meticulous study of how miasma spread cholera via stinking air?

And even if you can show a genuine health benefit from, for example, consuming less salt, individuals may decide that a very small reduction in their odds of a bad outcome, or a very small extension of their predicted lifespan, is not enough to justify giving up one of life's pleasures.

For example, as I write in 2015, newspaper headlines are proclaiming:

'Hot dogs, bacon and other processed meats cause cancer ...'
'Red Meat Cancer Risk To Be Revealed By World Health Organization'
'Processed meats as bad a cancer threat as smoking.'

Turning to Cancer Research UK (CRUK) for clarification, I read that the studies being reported show: 'those who ate the most processed meat had around a 17 per cent higher risk of developing bowel cancer, compared to those who ate the least.'

At this point, you may put down that half-eaten hot dog or bacon butty. But CRUK also points out that this is a relative risk. The odds of developing bowel cancer at some point in your life, across the whole UK population, are under 6 per cent. So increasing this risk by 17 per cent would raise it to ... about 6.5 per cent.

If 100 people all changed from a low-sausage to a high-sausage diet for the whole of their lives, one of them would probably develop bowel cancer who might not otherwise have done. We'd never know which person, and they're all hypothetical anyway, but you can see that one tasty meat product on one day is a small addition to your personal danger.

Red meat seems less risky, though it's still a suspect, especially if barbecued. Science can be cruel.

Across an entire population, if people gave up cooked breakfasts, salami and smoked ham as fast as smoking rates have fallen since 1950, bowel cancer rates would fall. Nowhere near as fast as lung cancer rates are falling, but enough to show in the data. CRUK estimates that, if nobody in the UK ate any processed or red meat, there would be 8,800 fewer cases of bowel cancer every year.[*]

---

[*] Compared to an estimated 64,500 fewer cases of lung cancer if nobody in the UK smoked.

However, for each individual, given that you are subject to all sorts of other risks in your life, and depending on how much you love bacon, it might not seem worth it. Big data is great at seeing the big picture, across a population. It's not so good at helping an individual make decisions about how they want to live their life.

CHAPTER SIX

Big society

L ast month,* 769 crimes were reported within a mile of my home. Among these, 186 were violent and sexual offences, 62 concerned drugs, another 62 were vehicle crime, there were 17 bicycle thefts and 142 were antisocial behaviour.

But I can be more specific. If I turn right out of our road, one of these vehicle crimes took place on the next corner. If I turn left instead, two of the violent/sexual offences happened just past the pub on the other corner. My usual route to the station passes the site of one violent/sexual, three vehicle crimes, one other theft, and one drug offence, plus three other drug offences at the station itself. And if you had to put up with the erratic train service we have here in Peckham, you might turn to drugs too.

I'm not just telling you this in the hope that property prices will temporarily dip enough for me to buy a flat here. I've looked up my local crime map, a service provided by the Metropolitan Police service online since 2011, and the slightly misleading precision of the flags on a map of my area is weirdly seductive. It's misleading because the locations are, for privacy reasons, unlikely to be more accurate than postcode or street name. So if your house happens to be the centre of a postcode, you may be appalled to find yourself at the epicentre of an apparent crime wave. Or delighted, if you're looking for a way to negotiate cheaper rent.

However, police ability to put reported crime on a map isn't just a quirky public service. It's being used to predict and prevent future crime. In spring 2012, a small group of British police arrived in Los Angeles, where the sunshine, golden beaches and

---

* Though by the time you read this, it'll be a couple of years ago.

the chance to rub shoulders with film stars must have made a pleasant change from patrolling the grey, rainy streets of Kent.

For those not familiar with Kent, it's a county in the south-east of England, stretching from London to the coast, that includes several ports and a tunnel linking the UK with France. Kent has Canterbury Cathedral, hops* and cigarette smuggling. It doesn't have Hollywood, palm trees or a murder rate of six per 100,000 people.

In fact, the UK recorded 551 murders altogether in 2013, fewer than the 600 murdered in LA County alone in 2012, giving us a national rate below one murder per 100,000 people. Nevertheless, the Kent Police delegation went to learn from the Los Angeles Police Department about one method in particular: predictive policing, and a data analysis tool called PredPol.

Thursday 13 February 2014 was such a good day for the LAPD that they issued a press release. In one district, Foothill, they had no serious crimes to report: no robberies, assaults, rapes or murders. Not even a burglary. Covering around 123km² (46 square miles), with a population close to 200,000 people, that district saw a 23 per cent fall in crimes reported over four years, from 2009 to 2013.

It's worth noting that crime is generally falling around the world, but even so the LAPD is proud of their recent record, which they ascribe to a mixture of working with local communities and applying PredPol to avert crimes before they happen. That's what brought Kent Police over in 2012. With their budgets being cut, could they use big data to apply shrinking resources to greater effect?

The theory of PredPol is startlingly simple. Using only the date, time and location of previously reported crimes, the software identifies 500 metre (just over 500 yards) squares at highest risk of crime within a specific police shift. So, instead of randomly

---

* A climbing plant used mainly to flavour beer, not a quaint one-legged mode of transport.

patrolling a given area, officers try to spend part of that shift inside these areas, or boxes.

Kent Police were impressed enough to give it a try. Over four months, they tested its predictive strength in one district, setting it against their traditional method, which was a human being spending several hours a day analysing crime reports and looking at maps. At the end of the trial, they said the software was over 50 per cent more accurate, as well as being quicker and cheaper.

Kent police are now using PredPol across their region, like many other police forces in the UK and US, including quite possibly mine. Next time I see a police patrol car pull into our side street, where the local youth like to sit around and chat of a summer evening, perhaps they're not just leaping to conclusions about what Peckham teenagers get up to. Perhaps, without knowing it, I'm living in a PredPol box where there is a high risk of a crime taking place.

If so, I wish they'd tell us. I'm sure that's information we could use. Which reminds me, I must look up the London data map of average rents for my borough and postcode.

Local data for local people

Average rents and local crime rates are not the only thing I can instantly look up about where I live. Right now, I can tell you the weather conditions in central London (cold and damp) and at London City Airport (mostly clear, 25.7kph (16mph) wind from the west).

All this, plus air pollution, live images from random traffic cameras, the FTSE index of stocks, and what's trending on Twitter in London, are displayed on a single dashboard at citydashboard.org, a website run by the CASA Research Lab at University College London. I can tell you that there's a good service on all Underground lines, that the Thames is 5.91m (19ft 4.7in) deep at Tower Pier, that 9,689 bicycles are available in Transport for London's bike-hire scheme, and that Londoners are 1 per cent happier than the whole country, and 1 per cent happier than our long-term average.

I wonder why we're reported as being 1 per cent happier than usual, so I follow the link to find the source of that figure. It's an app called Mappiness, run by a research project at LSE, The London School of Economics and Political Science.

Volunteers download the Mappiness app to their smartphone, and regularly report how they're feeling, along with what they're doing, where and with whom. They can also upload a photograph. While they're doing this, the smartphone's microphone will measure ambient noise levels, and the GPS will record approximate location. This information is aggregated, mapped and analysed to see if there's a relationship between how people are feeling and their environment at the time.

You can look up the results on a map, on which different coloured tags show where people have most recently reported being happy, or very, or extremely happy. Or you can see the hedonimeters, whose delightfully retro needles show UK and London happiness levels, and an 'average' needle for comparison, like a little analogue barometer of mood.

I say happiness: what they're actually measuring is 'momentary subjective well-being'. They text their volunteers and ask for an instant reading of mood. And if it sounds as if I'm being over-picky here, just try talking to a research psychologist about measuring happiness. Because the way you ask gets you very different answers indeed. Asking about life satisfaction gives you the impression that having children is the best path to a happy life. Asking about moment-to-moment mood tells you the opposite.

The Mappiness researchers are testing the hypothesis that happiness is greater in natural environments, and their initial data supports this claim. Which deepens the mystery of why people are happier in London than the UK generally, most of which is to some degree a natural environment, as you'll know if you've ever flown over it. Only about 3 per cent of the surface area is built on. And, more broadly, why worldwide we are moving to cities and abandoning the countryside in our millions.

That said, even in London more than half the space is either green or covered in water. So perhaps the happy Londoners are in parks, or walking along canals, or sitting in their gardens. Not at work, anyway.

I would download the app, thereby contributing my data to their research, but I have the wrong type of smartphone, so I can't. Which makes me a little unhappy. But there's no way they'll ever know that, or be able to include it in their data.

The Greater London Authority has its own city dashboard, with a more mundane selection of information: stuff like household waste, home repossessions, smoking quit rates and lost customer hours caused by delays on the Underground. Not so sexy for the casual browser, but more useful for governing a city.

In New York City, Mayor Bloomberg set up the Mayor's Office of Data Analytics in 2013, headed by a jovial man called Mike Flowers. Mike is nothing like the Silicon Valley nerd you might imagine. He looks as if he could play American Football, and he honed his data analysis techniques in Baghdad's Green Zone after US and allied forces overthrew Saddam Hussein's regime.

Nor did he start out with Stanford alumni and fancy software, just some keen young graduates working on spreadsheets like the ones you may have on your home computer for bills or football results. Nevertheless, by applying their analysis to the data already collected by different departments, they were able to make some big differences to how things were done.

The Fire Department, for example, would prefer to prevent fires, rather than risk their lives putting them out or rescuing people from blazing apartments. So they spend some of their time inspecting buildings, aiming to eliminate fire hazards before they turn into conflagrations. Historically, however, in spite of the best hunches of experienced fire officers, their success rate in finding risky buildings was little better than picking them out at random.

By combining the expert knowledge of the people on the ground with other data, Flowers's team was able to improve the Risk Based Inspection System. As well as the Fire Department's own records of previous fires, they pulled in information from the Department of Building, who also inspect potentially unsafe buildings, and other departments. For example, if the Department of Finance has an address listed as being behind with payments, it may also be behind with its safety measures.

Flowers has moved on, but the NYC Data Analytics office continues its work. And it's not the only organisation using data to get a better picture of life in New York City.

Making maps

I'm standing on a street corner in Brooklyn, surrounded by graffitied brick buildings and yards, frowning at my New York subway map, already battered and torn from all the times I've unfolded it the wrong way. As well as the spaghetti of coloured transit lines, it includes just a few key road names.

On the island of Manhattan, that's enough. The great thing about a grid of Streets crossing Manhattan roughly East to West, and Avenues running North–South, both numbered in order, is that it's very easy to work out how to get from station to street address. Most people will helpfully say, '23rd Street, between 7th and 8th Avenue'. Point yourself in the right direction, count the blocks, and you're there.

Here in Brooklyn, all that breaks down. Roads here just have names.

I know, I know, I should be using Google Maps on my smartphone. But I didn't want to pay my mobile provider's exorbitant roaming data charge, so I'm relying on this bit of folding paper. And yes, I'm aware of the hypocrisy of researching a book about big data while refusing to pay for data use.

There is a local street map in the subway station, which I thought I'd memorised well enough to get me where I'm going, a couple of blocks away. But now I can see the real streetscape, an apparently deserted jumble of industrial units and parking lots, the map I had in my head doesn't make sense any more.

A man who looks as baffled as I feel comes over to ask me for directions that I can't give. Forced to admit defeat, I go back down the subway stairs and translate the wall map into a mental list of directions. Left alongside the station, right at the end … I keep rehearsing the turning points as I walk between fashionably shabby lofts, hip new eateries and apparently empty warehouses.

One of which houses my destination, CartoDB, where they use big data to make maps.

They've just moved into this place, a sleek conversion that's now all steel, glass and the compulsory polished-concrete floor. Outside, the graffiti covering the parking lot walls is so gorgeous that I ask who did it for them. It was there before they moved in, they say, but I think they plan to keep it.

I'm visiting Stuart Lynn, a map scientist, which, as he says, 'is a pretty good job title. It's better than geographer, which I feel quite good about, because after leaving astronomy I wanted to still think of myself as a scientist.'

He's a cheeky, self-deprecating Scot, though his accent's veering towards Canada, possibly the effect of a few years working in Chicago on a project called Zooniverse. That enlists citizen scientists to help analyse the flood of data pouring into research projects, from astronomy to zoology.

Stuart is surprisingly sympathetic to my story about not being able to find my way to a mapping company. He says he finds navigating around New York confusing after Chicago, where the streets have numbers but the avenues have names. It's data he loves, not maps. But Stuart is genuinely excited about all the things you can do when you matchmake between the two.

Join up two public databases and predict whether a house has a working smoke alarm, for example. Or produce a 'heat map' of public response to Presidential candidates, using Twitter mentions to colour-code the areas with the strongest response to Trump or Sanders. Stuart compares that to taking spot temperature measurements outside, and using them to model a continuous field of temperature, 'so you can think about tweets as being a sentiment field for a politician'.

On a less serious note, he shows me a map of 'reports of human waste in San Francisco'. I'm disappointed to report that the blobs on the map are blue, not brown.

Why is using maps to visualise big data so popular?

'It parallels what I think happened with maps when people first started to use them,' says Stuart. At first, they were just a way of conveying where things are, and some ideas about what might be there, but nobody was sure. He jokes about a 'sea monster probability field', not an absolute guarantee of meeting a sea monster, but an increased likelihood.

I'm reminded of Galton's use of maps to develop his understanding of weather, combining wind speed and direction to see the emerging pattern of the anticyclone. Though his maps were more sophisticated than 'here be monsters', and based on empirical observations.

'But then you see this shift around the First World War,' says Stuart, 'when people begin to realise that maps can be a tool either for making your point more clearly or obscuring your point. A map is a political tool. I think that's when people learned that maps were a good way to visualise data. You could tell a story with them.'

And that's what excites Stuart and his colleagues at CartoDB. 'We feel very strongly about maps and geographical data because we think there's a huge amount of power in them for individuals and for small organisations. The map makers were always governments, but we think that with a lot of data from the government being open now, you can grab data from Chicago and New York, you can find out crime statistics, all that sort of stuff …'

At this point we're interrupted by a small dog chasing a couple of even smaller drones, being flown down from the mezzanine floor by one of Stuart's colleagues. Honestly, sometimes I wonder if these companies stage things for my benefit, to prove just how tech startup they are. A little terrier alternately pursuing and running away from a pair of drones the size of large moths could be some kind of metaphor for our relationship with technology, but it isn't, it was just a funny diversion, though I did feel a bit sorry for the dog.

Back to the main point. Stuart Lynn's job is mostly about working with small organisations or individuals, including journalists. CartoDB aims to give them tools to quickly and easily turn data into maps.

'We embraced this idea of a citizen cartographer. Someone who is collecting data themselves, or using data that is open from the government, to make maps that are going to impact their communities, their countries and their peers,' he says.

'Trying to move beyond that, the next phase of what we're doing is to democratise location intelligence. How can you use our tools to get a new sense of the world? Not just visualising, but being able to call on statistics and some basic techniques to really extract information.'

He gives me an example: 'Here's a bunch of questionnaire responses from people in your local neighbourhood – how does that compare to the national average? Is your neighbourhood more likely to have diabetes than other neighbourhoods in your area? Those kind of questions. That's not easy though, that's the next big challenge I think.'

If you come from astrophysics I don't suppose 'easy' is the first thing you're looking for.

'Weirdly though, people are often more difficult than stars and galaxies.'

I would hope so, I say. If I was as easy to understand as a star or galaxy I would feel I was failing as a human being. I aspire to at least being dark matter.

Stuart Lynn laughs.

'When we were working with Zooniverse, we worked with some sociologists at Syracuse University to understand the motivations of citizen scientists.

The first time we went through this with them, we were brainstorming for two days: things we could do, experiments we could run ... and there was this pattern formed. They would suggest this theory we would want to test, I would be like: We could take this data and that would give us a heuristic to tell us whether people were doing it because of this ...' Stuart waves his hand in illustration.

'And we would go on for five minutes, and then one sociologist would say: "Well, we could just ask them."'

He laughs, 'And it hadn't occurred to me that was an option. Which is a thing that you get when you study things that are very far away for a long time.'

You can't ask a galaxy why it does what it does. But you can ask a human being. They may not tell you the truth, or even fully know themselves, but it's a start.

Stuart shows me a map on his laptop screen. At least, it looks like a map, with blocks separated by dark bands that are probably roads, but it's only made up of coloured dots.

'This is a map that has no base map at all,' he says, ' this is just data. This is Chicago, this is the river,' he indicates a broader black line, 'and every point you see here is a crime.'

And the crimes are colour-coded, I can see now. He goes on, dissecting the anatomy of the city that was his home until a few weeks ago. 'Very, very quickly you see this huge discrepancy between these two areas. This is downtown where all these crimes were credit card fraud, financial identity theft ... and then not too far away is this area.' Mainly crack, heroin and cocaine, says the map. 'A much, much poorer area, and for anyone who knows Chicago this will jump out, they will know exactly where that is.'

Being able to zoom out like this, and see the patterns underlying each individual crime, is very useful if you want to do something about reducing crime in your city. But it makes me uneasy. If you live in Chicago, your address or neighbourhood now tells me something about how much crime happens around you, and what kind. It would be like my address ending, 'Peckham, drugs and bicycle theft, London, England.'

If you're the Chicago Police Department, or London's Metropolitan Police Service, you probably already know where crimes tend to happen, even without a predictive computer model. But any model would be based on the past data. You can only predict the future if the future behaves like the past. Sophisticated models can respond to change, and update themselves, but by definition one can only gather data from the present and past, not the future.

So might using data to make public maps of where crime happens, and what kind, become a kind of prophecy for the people who live there? A self-fulfilling prophecy, even?

'Yes,' says Stuart Lynn, 'I think that's hugely important. I think that's a common problem with a lot of social advocacy. Once you name a problem it becomes self-stigmatising.'

Part of the point of his work at CartoDB is to put that map-making power into the hands of communities, to tell their own stories, to use data for their own purposes. He recalls a program in Chicago called: 'Data Science for Social Good, and there was an interesting project trying to make a predictive model of whether kids in high school were going to graduate. Or if they were going to graduate late, because graduating late is almost as bad as not graduating at all, it can have a huge knock-on effect on people's future careers and life,' Lynn says.

'And so they were trying to predict, given the data that they had for those kids each year, whether or not there was a high probability of them baling out of school and whether or not there should be an intervention to help them. Which is amazing, great, but they were doing it at classroom level, school level, district level, and also on an individual level, trying to get the probability for the individual kids about whether they were likely to graduate.'

So far, so good. But as Stuart Lynn points out, the cost of a false positive, a kid who is doing fine, but gets flagged up as at risk of not graduating, could be high. Suddenly they're the focus of unneeded attention, they're self-conscious, perhaps their confidence is knocked.

'My worry about all this stuff is you end up where there is no recourse and no person. I think projects like that where you could flag something up to the teacher, and they could say, "No, this kid is okay", is where the balance comes in. It's similar with maps as well. I think if you have the data on the map and look at it, the first thing you do is say: Ha! that's interesting, I didn't expect that, I wonder why that is? And you interpret it.'

He relates it back to his experience at Zooniverse.

'I think this is one of the things we saw in the science projects, the human judgement in that loop. You find things you would have missed, you find planets you would have missed. There's huge value to having at least some human intelligence in that system. I think that's probably always going to be true.'

Taxi to nowhere

Stuart shows me how CartoDB's mapping tool works. First, you simply put your data, or a dataset from their library of data, into the system. The New York City taxi data* for example.

'You can see a table of the data. If there's a column called latitude and longitude or something like that, it'll automatically detect it and put it on a map. Or we could do city names, postal codes or street addresses, through a number of different geocoders. Once you've got it there you can see the map.'

And there, indeed, is the New York City Taxi map, overlaid on a world map. Mostly in New York, as you would

---

* Chris Whong, who first obtained the data with a Freedom of Information request, now works for CartoDB. He spotted a tweet that included a graph of daily cycles of taxi availability, asked for the dataset, and was told to fill in a form and then bring a 200GB, unused hard drive into the office. Being an Open Data enthusiast, he then made a version available for public download.

expect. But a few points clustered off the west coast of Africa, somewhere in the Gulf of Guinea.

They must have tipped well.

'There's this fun thing in mapping,' says Stuart, 'where sometimes your data is going to be off, and you lose latitude and longitude. There's this little imaginary island that we like to think of at 0.0° 0.0°, null null, called Null Island. Everything that gets lost in the world goes there.

So if you can't geocode something it gets a latitude and longitude of null, which means it goes to point zero zero on the map.' He points to the stray blobs in the Gulf of Guinea. 'So these taxis are on Null Island.'

For a moment I imagine a couple of yellow taxis on a sandy beach, half buried in a massive pile of my odd socks.

'But the bulk of these are here in New York.' Stuart zooms in on the city. 'So here's all the points. We can cluster them, we can colour them differently for different categories. Here's a map of drop-offs and pickups. As you'd expect, a lot of pickups in Manhattan and drop-offs out-of-town, and that's the airport,' he points.

'With paper maps you can never really show changes over time. You can do that online really easily. You can have this visualisation of how pickups and drop-offs vary over time. Now you can see some of the rhythms of the city, day in, day out.'

As the days flick by, the map pulses in red and blue, like the city's heartbeat.

I love maps. I could sit and look at them all day, even on paper. On screen, pulsing and evolving and zooming out to see the big picture, I could be here all night. But it's the end of CartoDB's first day in their new building, so there is beer to be drunk, and a small, baffled dog to be comforted, and I am warmly invited to join in.

And then I have a trivia night* to go on to, so I reciprocate Stuart's hospitality and recruit him and his partner to the

---

* Pub quiz, British readers.

team. This time, I leave the journey planning to him, since he does have mobile data, and is a professional map scientist.

I look up my directions. It's a bar in Queens, on the corner of 24th Avenue and 37th Street.* We change subway lines, emerge from another station and walk up the road as Stuart's smartphone instructs.

Brooklyn is newly gentrified, but this part of Queens is still an endless strip of car-repair workshops and valet-cleaning places, the water sprays diffusing red and yellow neon light against the night. There's a mysterious smell of pear drops that vanishes as we turn a corner. I suggest a smell-map of a city, and Stuart tells me it's been done: CartoDB have a stink-map of London.

We turn off the main road and traverse the couple of blocks to 24th and 37th. But there's nothing there. Not even a car-repair place. Just locked-up shops and small apartment blocks. I check the directions, in case this place is around the back of something. That kind of knowledge isn't always there to read from a map, however hi-tech. Sometimes you need a human for the detail.

Hang on, I say, is this the corner of 24th Avenue and 37th Street? We all look at the sign.

It isn't. We're at the junction of 24th Street and 37th Avenue. Thirteen blocks from the bar in both directions, north and east. Twenty six blocks to walk. Curse those numbered avenues of New York.

Stuart calls an Uber, which arrives in about two minutes. I guess the driver has satnav.

Citizen data

Worldwide, under the snappy 'smart cities' catchphrase, local governments are looking to the plethora of data that their populations generate, and wondering how to use it. Cities

---

* Manhattan doesn't have that many avenues, so we can't make that novice mistake.

including Pittsburgh PA, Kyoto in Japan and Bristol in England are experimenting with infrastructure that depends as much on data as on physical construction.

Much of the smart city agenda is about energy use. One Danish suburb is trialling streetlights that only light up when they sense somebody approaching. In Kyoto, one project experimented with explicitly reducing energy use within the home simply by showing each household a visual display of their current usage, combined with other methods such as sending messages asking them to use less, or offering points that could be exchanged for money.

Glasgow won a UK-wide competition for £24 million funding to become a Future Cities Demonstrator. They plan to use the hundreds of data sources already available to them on everything from school electricity use to rubbish bins. They already have their eye on a more efficient use of street lighting, along the same lines as the Danish technology.

They're also keen to involve the population in collecting and using data, and not only by getting walkers and cyclists to share route information using smartphone apps. In a series of hackathons, intensive brainstorming events for people with data analysis and other tech skills, they awarded prizes for ideas to use the city's data. The energy-themed event was won by a team that will use their £20,000 prize to develop their concept: an alert that tells council employees when energy use in their building is reaching a certain level so they can go and turn off some things.

You can see the city's data at open.glasgow.gov.uk. As I write, there's a limited range of information, including cycle parking and the location of CCTV cameras.* Part of Glasgow's Future Cities project is a new network of 400 digital cameras,

---

*Readers outside the UK may be confused by the matter-of-fact way I describe the routine surveillance of our public spaces by cameras. With between 5 and 6 million cameras in the UK, we have one per 10–15 citizens, making us the most-watched nation on Earth. Still top at something! Rule Britannia.

with a brand new control centre, to replace the old CCTV cameras in the city's public spaces, and connect them with the traffic cameras and traffic light controls.

Which makes sense, of course, if you want to use the information you have coming in. Putting together different sources of data to get more than the sum of the parts is a key part of big data's promise. But to me, at least, it also feels a little Big Brother.

One control centre from which the council can survey the city, control street lights and traffic lights, compile information coming in from buses and smartphones* is a wonderful thing if you want to keep traffic flowing smoothly, or respond to an emergency. But it's also a wonderful thing if you wanted to spot, control or disrupt a peaceful protest.

The capacity to gather information, draw conclusions from it and apply it to public policy can be used for different purposes. And sometimes, you may not agree with those purposes. Or, even if you do agree with those purposes, you may feel that the measures taken are disproportionate to the problem.

The UK Government routinely collects so much data on its citizens through different branches that its Office of National Statistics (ONS) suggested abandoning the 10-yearly census. Between taxes, the electoral roll, National Health Service records and specific services from driving licences to schools, the state has plenty of basic information about each of its citizens without going to the expense of sending out census forms and uploading the data into yet another computer system.

By today's standards, a dataset updated every 10 years, that takes months to gather and process, is lumbering and old-fashioned. As Stuart Lynn pointed out, the Brooklyn neighbourhood where CartoDB relocated had gentrified within a couple of years. Between one census and the next, an area can shift from low-rent apartments and car repair shops to hi-tech startups and expensive pizza joints.

---

* Free open-air Wi-Fi has its price.

The ONS suggested that they might supplement administrative data with annual surveys of 4 per cent of households, samples that would be quicker to process and give a good enough picture of what we're thinking or doing. But although the idea of more frequent information was welcomed, public consultation convinced the ONS to hang on to the decennial census for now, though it will move online in 2021, and will probably change beyond recognition in the next few decades.

One of the reasons cited for keeping the UK census was concern that data given for one reason might be used for another purpose, or passed to third parties, without consent. That fear might lead people to give false answers, rendering the information less reliable.

The UK census of 1991 lost a 'missing million' of people who were not counted. This was linked in news reports at the time to the introduction of the unpopular Poll Tax, a new form of local tax charged to each individual.* Those who planned not to pay the Poll Tax, in protest or to save the money, feared they would be identified from the census return and end up in court.

Keeping the census separate from other data was therefore seen as part of a relationship of trust and willing participation in gathering information.

However, regardless of the future of the census, local and national governments are collecting immense amounts of data on us all. Most of this data will be used to plan the roads, railways, schools, homes and hospitals of the future, something from which we all stand to benefit.

There's also some potential for keeping an eye on your behaviour, from unhealthy lifestyle habits to antisocial tendencies.

---

* Previously, a tax was levied on each property, varying with the size and value of the property. The new Community Charge, to give its official name, was levied on each person. This was widely seen as a regressive tax, as it took little account of variations in wealth.

At risk

Graeme Tiffany is a Community Philosopher in Leeds, in the north of England. I met him at a packed Leeds Salon evening, where we discussed big data with a lively and critical audience. Graeme introduced us to his thesis that there is a new prejudice at work in public policy, one more acceptable today than racism, sexism or even talking about somebody's social class.

For Graeme, the pernicious idea most likely to blight the lives of young people before they've even got started is 'at-risk-ism'.

It was reading the UK Government's 2008 Youth Crime Action Plan that first made him feel uneasy. On page 1 he read that half of all youth crime is committed by around 5 per cent of young people, and that:

'Increasingly we know how to identify these young people early on – in particular how they tend to come from a small number of vulnerable families with complex problems … We will offer non-negotiable intervention to the families at greatest risk of serious offending.'

The document goes on to list risk factors, including low IQ in child or mother, a parent with a criminal conviction, low socio-economic status, maltreatment or a diagnosis of attention deficit hyperactivity disorder (ADHD). Between 20 per cent and 40 per cent of high-rate offenders have one or more of these risk factors in childhood.

You'll have spotted, of course, that saying '40 per cent of high-rate offenders have low socio-economic status' is not at all the same as '40 per cent of people with low socio-economic status go on to become high-rate offenders'. Thankfully. That truly would be a crime wave worth worrying about. It's a common confusion, akin to saying that since 90 per cent of murderers are men[*] then 90 per cent of men are murderers.

---

[*] Men are also twice as likely as women to be the victims of murder.

Nevertheless, and noting that the same factors are also associated with low educational attainment, ill health and disaffection, the strategy document outlines an approach that includes providing more youth facilities and support services for families.

This isn't new, of course. Quetelet published his figures on crime in 1831, not to show off his mathematical prowess, but to demonstrate a wider point: that crime occurs in regular patterns, suggesting that it is subject to forces wider than individual moral reasoning. If there is more to crime than a few bad people, or people making bad decisions, then it should be possible to reduce crime by working on those wider factors.

If you look back on your own life, and decisions you made, good and bad, consciously or without knowing till later that you'd made them, you can probably see points where circumstances swayed you one way or another. Perhaps the lack of a certain subject at your school pushed you on to a different career path, or you went along with the interests and habits of your friends.

Perhaps you happened to work with somebody who was obsessed with the flying trapeze, and somehow that led to you touring with a circus for a few months and then becoming a comedian. Hey, we've all done it. My general point is that trying to change people's behaviour by changing their circumstances is not a ridiculous approach.

You can also argue that providing better facilities for young people, especially in poorer areas, is a good thing in itself. However, policies today need to show evidence, preferably numerical evidence, that they will help achieve certain outcomes. So a bar chart showing that more than 40 per cent of high-rate offenders have low socio-economic status is much more likely to get you funding for your adventure playground or whatever.

Graeme Tiffany has been involved in youth work for a long time, and probably has no objection to spending money on some facilities for young people, though he might prefer to offer them philosophy clubs. He's less happy about the proposal to 'offer non-negotiable support' to troubled families.

As he points out, 'non-negotiable support' means state intervention into family life, regardless of consent. And in many cases, as this is a preventive strategy, this will be state intervention into families where no crime has been committed.

You may, reasonably, be thinking that it's better for all of us, for wider society and for the potential young criminal, to act before things go wrong. But imagine a slightly different scenario. Imagine that one of the risk factors for being a young criminal was being from a certain ethnic group: Black British, or African-American, for example. Imagine that the data shows 20–40 per cent of young criminals are from this ethnic group and therefore, to protect them and the rest of us, families from this ethnic group will be offered 'non-negotiable support'.

In the 1980s, riots in several British cities were sparked by hostile relations between police forces and local people. The police were using so-called 'sus' laws to stop and search people on suspicion of intent to commit a crime. Most of the people stopped and searched were young Black men.

The police at the time would have claimed that this was not prejudice, but recognition that most of the perpetrators of certain crimes, such as mugging, were young Black men. More than 40 per cent of police searches under the sus laws were carried out on Black people, but they only made up 12 per cent of all arrests. Which suggests that police perception of likelihood of guilt outweighed actual likelihood of guilt.

From the point of view of an individual who may be stopped many times in a week purely on the basis of a simple profile – young, male, Black – it was pure prejudice. Literally, being prejudged. The opposite of innocent until proven guilty.

Returning to our families *at risk* of producing young criminals, you can see why Graeme talks about 'At-Risk-ism'. A few factors, about which you can do nothing, flag you up as a potential criminal, and suddenly your parents are under scrutiny from the state. For your own good, of course.

Intervening in the families of potential criminals is not a particularly big data approach in itself. But using risk factors to decide which families to target relies on linking different datasets: school records, low-income status or previous police interactions.

But whether it's based on crime maps, computer models, or on the hunches of local police forces, the idea that some people are measurably *at risk* of a bad outcome is one that we'll return to later.

CHAPTER SEVEN
Data-driven democracy

I like to experience new food and cuisines, and to surround myself with a diverse range of cultures and ideas, but I find the idea of being in debt stressful. I am younger than the average UK voter, more libertarian, less nationalistic and more Green. I get my news mainly online from Facebook, from the BBC, and from the London *Evening Standard* newspaper. My favourite celebrities are Angelina Jolie and Helen Mirren, and my favourite TV viewing is sport, specifically, the recent Commonwealth Games and soccer.

Some of that is true. I probably am more libertarian than the average UK voter, and I do like new foods and diverse cultures, which is one of the reasons I chose to live in Peckham, an ethnically diverse and fast-changing neighbourhood in south-east London. But I'm not very Green, I'm indifferent to Angelina Jolie, and I watch sport only so I can check whether it's a good day to call my dad.*

The profile in the first paragraph is not me, it's the average voter in my constituency according to Pollsters YouGov. The UK's 2015 parliamentary election took place as I started writing this chapter, and if you're starting to think: 'Oh yawn, politics is so boring!' *don't* skip straight on to Chapter 8. We're going to look at why politics is so boring these days, and even some ways we might restore it to life.

YouGov predicted how each UK constituency would vote in 2015, and not just by asking a sample of people how they planned to vote.

---

* If Liverpool FC are winning, it's a good day. If they just lost a match, I might leave it for 24 hours.

YouGov regularly interviews thousands of people across the country, about all sorts of topics. So their profile of a constituency won't just tell you how people say they will vote, or some relevant information such as how rich they are, or how young. It will tell you where they stand on a range of political topics, from abortion to nuclear energy, even how likely they are to watch X Factor or to take off their make-up before going to bed.

'A political campaign is the ultimate marketing campaign,' say YouGov, and, 'at a time when MPs are seen as being more "out of touch" than ever before, and confidence in the political process is at an all-time low, we think that giving candidates some hard data about what their constituents are really like can only help the democratic process.'

All the UK's major political parties have been using this kind of technique for some time, though not with such sophisticated methods for collecting and analysing the data. The Labour Party has a clever interactive app called *how many of me?*, which offers to tell you how many other UK voters have the same name as you.

It also collects your email address and tells you that the Labour Party or its elected representatives may contact you 'about issues we think you may be interested in or with campaign updates.' But that's in smaller print. I'm pretty sure that I am the only Timandra Harkness on the electoral register, so I didn't bother to give them my email address.

In the previous UK General Election in 2010, the Conservative Party used a database bought from a credit reference agency, Experian. The Conservatives put the data into their own system and used it to target people according to their consumer tribe. So if you were a busy young parent, you might have received a leaflet pointing to help with childcare, the quality of the local schools, and so on. Retired people might have been given a different leaflet with more about pensions, healthcare and crime.*

---

* I'm making these up, for all I know there's some research predicting that pensioners respond best to pictures of sports cars, and young parents to reducing tax on beer.

This time, all the parties upped their game, stealing ideas, software and even people from Barack Obama's successful election campaign. For example, LA-based NationBuilder provide a software system that can keep track of people who have given you their email address and link to their Twitter, Facebook or other social media profiles. That way, you can see what they're posting online and target them accordingly.

Political campaigners, just like the people who want to sell you things, can now know more about you than whatever you choose to tell them. You may tell every candidate who knocks on your door that you plan to vote for them, so you can get back to watching X Factor. Once, they'd have to take your word for that, unless you'd also put a poster for the opposing party in your window.

Now, they may know that you mentioned on Facebook your plan to vote for the other candidate, because of their enlightened policy on beer taxation. Or that you tweeted you were undecided, and that all your other tweets are about how bad the local train service is, or how worried you are about crime, ever since you looked at the crime map of your street. So to win your vote, they need to reassure you that beer tax and crime will fall, and train services improve, once they're elected.

But why are political campaigns so keen to gather all your data?

Partly because they can, of course. For a few weeks after I visited the website of NationBuilder to find out what they do, promoted tweets from them appeared regularly in my Twitter timeline. Which is a bit creepy, since I didn't even give them my email address. All I did was visit their site and look at a few web pages. I tremble to imagine what they could do if I actually gave them some information.

If NationBuilder can pop into my Twitter timeline simply because I look at their website, imagine how easy it is to put together a few databases – the electoral register, social media sites and credit reference agencies, for example – and get a snapshot of me and all my neighbours.

The software to look for patterns is readily available too. As part of their 'how can we stop people getting bored with this election?' coverage, the BBC commissioned a study of how names are connected to your voting tendencies. They found that Nigels are roughly twice as likely as the population in general to vote for UKIP, the party led by Nigel Farage, but that going on her name alone, the Conservative leader's wife, Samantha Cameron, was more likely to vote Liberal Democrat. Though presumably she wouldn't be voting on her name alone.

By this point you may be thinking: So what? Political parties now have a more efficient system for listing potential voters, but is that really a transformation of how democracy works?

That's a fair point, thanks for making it. Your opinion is valuable to me. And when I have fed it into my software I'll be able to assess whether millions of potential readers feel the same way, and whether rewriting this chapter for the next edition might persuade some of them to buy the book and get me to number one in some bestseller list.*

Let me try to convince you that data-driven democracy is something more than a catchy headline or a more efficient filing system.

We don't talk any more

For many years, Britain was a two-party democracy, leavened by smaller parties who had a voice in parliament, but dominated by the Labour Party and the Conservative Party. Most people who voted had a long-standing allegiance, probably based on family loyalty. Working-class voters tended to feel that Labour represented their interests, while people who owned businesses were more likely to feel the Conservatives should be running the country.

---

* Some very niche bestseller list, admittedly.

Today, these loyalties are breaking down. Membership of political parties has fallen, like membership of other social organisations, from trades unions to tennis clubs. It's no longer enough to assume that certain groups of voters will be on your side by default. In fact, it's hard to assume anything, as you probably have no contact with potential voters from election to election, outside the few who are still active party members, and the individuals who want your help with some problem.

So how are politicians to gain the support of the electorate? How can a political candidate know what makes us put a cross in a box, let alone what makes us tick?

The best way is a face-to-face conversation, but you don't have time to do that with everybody. Some of us will never vote for you, so it's pointless to spend campaign time arguing with a staunch opponent. Some of us will vote for you no matter what, so, though it seems ungrateful, there's not much point knocking on those doors either. The important people are the ones still to be won over.

That's where data can help. If you can reach the potential-but-not-in-the-bag voters, you can change the outcome of an election. If you can reach them in the most effective way, by identifying the issues on which they might prefer you to your opponent, all the better. And if you can contact them directly through social media, direct email or by letter, better still.

Which is exactly what the Conservative party did in the 2015 UK general election, targeting just 80 marginal seats and identifying very detailed groups of winnable voters within those constituencies. Then they identified individual voters who might be persuaded to vote Conservative, and sent them personalised letters or messages that addressed their individual concerns.

Earlier still, their use of polling and focus groups had suggested areas on which the electorate as a whole were more inclined to trust the Conservatives, such as the economy or not being bullied by the SNP.* So the overall campaign was based on feeding back to voters their own concerns, with policies designed to offer convincing answers.

The Conservatives commissioned their own polling, not made public, which predicted overall victory for their party, in contradiction of all the polls predicting another hung parliament. In fact, it now looks likely that they deliberately allowed the media to report a neck-and-neck race when they knew they would be several lengths ahead, to help motivate voters to get out and cast their vote on the day.

If you feel a bit sad that politics seems to have changed from a clash of ideas about what kind of world we want to live in to a marketing problem, I'm with you. Choosing a government shouldn't be the same as choosing a new car.

Many of those involved in the campaigns would claim that politics is still about ideas, principles, visions of the future. The campaign team that took the Conservative party to victory ascribe it not just to data, but to a better leader, better policies, a disciplined campaign and a tightly focused message.

But politicians also want to get enough votes to put them in power so they can apply those principles. So they address their campaign, their manifesto, their headline policies, to specific groups of voters who can make the difference between winning and losing an election.

No wonder we're all so bored by politics. What should be one of the most important decisions we make, deciding the direction we want our country to take for the next few years, is reduced to picking the package that best fits our needs as a consumer. What's worse, it's a package tailored to what the data predicts we already want, like online shopping.

'If you like this policy on crime, you may also like these policies on transportation and beer.'

---

* Scottish National Party who, as predicted, won most of the seats in Scotland, and became the third largest party in the UK Parliament.

So much for what politicians know about us voters. What about the other way around? We can scrutinise their past record like never before.

Watching you watching us

In the UK, a website called theyworkforyou.com lets you put in your postcode and read your MP's record. I learn that my Member of Parliament has asked a number of parliamentary questions about disruptions to train services in our area, accepted modest financial support from a trades union, and a donation of two tickets for the Royal Box at Wimbledon from the All England Lawn Tennis and Croquet Club Ltd. If I cared to, I could look up what expenses she's claimed.

The basic information so readily searchable on this user-friendly website comes from official parliamentary records, so it was available all along to those with the time and know-how to search through Hansard and other paper records. Since the site has been compiling its measures, some enterprising researchers have written extra questions for their MP to ask, knowing it will improve their ranking and apparent diligence. That's the problem with measuring human behaviour: unlike galaxies or mosquitoes, we may change our behaviour to game the results.

On a page titled 'Numerology', I learn that my MP spoke in a below-average number of debates, but voted more often than average. The Numerology page carries the disclaimer: 'please note that numbers do not measure quality'.

In fact, elsewhere on the site it urges the user: 'when you're judging your MP, read some of their speeches, check out their website, even go to a local meeting and ask them a question. Use TheyWorkForYou as a gateway, rather than a simple place to find a number measuring competence.'

Wise words.

So can big data help us make properly informed decisions about who to vote for?

What do we want? Data! When do we want it? In real time!

A few months before the UK General Election in 2015, the Royal Statistical Society (RSS)* published a Data Manifesto.

If you're surprised to find an organisation dedicated to the technical manipulation of numbers getting stuck into politics, on a strictly non-partisan basis, remember that from the start it saw its mission as the improvement of society. Their very first meeting, when they were still called the Statistical Society of London, passed a motion stating that: 'accurate knowledge of the actual conditions and prospects of society is an object of great national importance not to be obtained without a careful collection and consideration of the facts.'

The 2015 Data Manifesto has three sections: prosperity, policymaking, and *data to strengthen democracy and trust,* and boldly claims: 'What steam was to the 19th century, and oil has been to the 20th, data is to the 21st.'

Data as fuel, again.

So how does the RSS think we can best use all this data? One way is to make existing data more available to more people. Not just government data, but commercial and scientific data. Partly, this is sheer efficiency. Sharing data can be quicker and cheaper, and lead to better-informed decisions.

But there's a less pragmatic aspiration, to strengthen democracy and trust. Instead of data being something that's collected about us by companies, governments and scientific researchers, why can't we all have access to it?

This is the idea behind 'Open Data', a systematic making available of data to anyone who might want to use it. People involved in the situation may have some insights that government statisticians, in their far-away offices, can never have. Simply by bringing many minds to bear on a problem,

---

* The same one founded in 1834 by Charles Babbage and his fellow enthusiasts.

we can come up with more ideas, some of which lead to a better solution.

And we're better equipped to test and challenge the claims of official bodies if we also have access to their raw material. Stuart Lynn and his colleagues at CartoDB are very driven by this ideal of putting the data, and with it the power, into the hands of ordinary people, letting them tell their own stories and define their own problems.

Though there are some practical hurdles before you can just throw all your figures online.

How to be open

Your data needs to be in a format that others can readily use, not just a photo of some figures scribbled on paper, like the one in front of me that tracks words written, hours spent pounding a keyboard, and cups of tea drunk.* I've heard many stories about Freedom of Information requests being answered with paper copies of tables of figures, requiring hours of human time before a computer could make any use of the information.

You need to state that you give permission for others to use and share it, and make this easy and practical. You need to make it traceable back to you, and to the process that produced it, so people can check it.

And, of course, you need to be sure that making it public won't infringe upon anybody else's privacy. Which is not as simple as just removing names and addresses.

Data scientists and statisticians have tools to help anonymise or pseudonymise – pseudo-anonymise – datasets. For example, suppose you're the only ethnic-minority family in a small Welsh village, with two kids at the village school.

---

* 79★ so far, but I'm only on the first draft.

   ★ 219★★ on second draft.

   ★★ 246 on draft four. But I'm increasingly going over to whisky since draft three …

Since the Department for Education keeps track of school students' characteristics, including ethnic background, gender and age, I could look at the test records for that school, classified by ethnicity, and see what marks your kids got, couldn't I?

No, I couldn't, because before those figures are released to the public they're adjusted to take out such revealing individual results. This is quite a complicated business, because it's important not to skew the overall figures too much, so taking out one entry means adjusting some other datasets too. But it can be done.

However, one of the great opportunities of big data, the ability to combine different datasets, is also a threat. Our old favourite, the NYC taxi database, was released through Freedom of Information. It's since been used to track where celebrities went after leaving public events, and to imply that they failed to tip. Which may have been unjust, as cash tips don't show up on the database. Frequent visitors to strip clubs or gay nightclubs could also be traced, at least as far as their home addresses.

In January 2015, researchers at the Massachusetts Institute of Technology (MIT)[*] showed how easy it is to identify individuals from a few pieces of information. They started with the credit card records of 1.1 million people, with all identifying information removed. All they had was the date and location of the transactions.

Using a few extra clues, such as tweets about eating out, or Instagram pictures of new clothes, Yves-Alexandre de Montjoye and his team found that just four transaction records were enough to identify the person whose credit card record they were reading, 90 per cent of the time. If they also had approximate values for the credit card spending, only three purchases would identify 94 per cent of shoppers.

---

[*] UK readers, don't be thinking of MIT as a technical college, where local teenagers go to learn plumbing and hairdressing. It's one of the most prestigious science universities in the world.

Two years earlier, the team had done a similar exercise with cellphone records that gave very similar results. It's never been easier to link datasets, including many that we make public ourselves, such as social media.

This loss of privacy is something I'll come back to in the next chapter. For now, just bear in mind that one person's open, transparent, democratising data may be another person's personal life.

Collective wisdom

So let's suppose that you have suitably anonymised your dataset, put it into a form that is readable by other people's computers, and made it easy to find and download. How can it be used?

One way is simply to invite anybody and everybody to pitch into solving a problem. Hackathons are endurance events, like a triathlon for people who prefer coding to cycling, swiping to swimming, and running simulations to running marathons.* They bring together people with technical and other skills to address problems, come up with ideas and start creating the computer code to make them happen.

They don't have to include free pizza and dozens of people who don't leave the room for 48 hours except to pee, but they often do.† Hackathons can be competitions, and are a popular way for the tech companies to find new ideas and talented individuals, but they can also be cooperative events with a common social aim.

---

* This is a bit unfair. Many of them do both, especially in California. They use apps like Strava to turn exercise into yet more data.
† If you think that sounds like what you did at college when you had a deadline to meet, you're right. That's how some of today's biggest technology companies started out, with a handful of smart young people pulling all-nighters over a computer.

San Francisco is a city regularly shaken by earthquakes, like the rest of the Bay Area. The Bay Area also includes Silicon Valley, the universities of Stanford and Berkeley, and the city and port of Oakland, so it's home to a world-leading concentration of technical know-how and entrepreneurial dynamism. Among San Francisco's many hackathons of 2014 was one to come up with ideas that would help after an earthquake or similar natural disaster.

Six months on, I gatecrashed the follow-up meeting to find out how the ideas developed so fast were taking shape. One group had created a cellphone app that could upload reports or photographs of earthquake damage to a central map, accessible by emergency services and the general public. Another had made a database where anyone can post offers of blankets, food or whatever, and connect their offers with requests for practical help.

And, because all these apps assume you have your cellphone or computer working and fully charged, a third group was creating a network of solar-powered information posts that double as charging stations. Prototypes were already being installed on university sites around the US, as you don't need an earthquake to appreciate a place to hang around, shaded from the sun or rain, and charge your phone for free. In fact, it'll work much better in an emergency if you're already in the habit of using it, and know where it is.

Running the meeting was Professor Dirk Helbing, a man who stood out against Silicon Valley's uniform of jeans and T-shirts by wearing a striking grey suit and tie, his hair cut into a severe fringe that out-nerded the nerds so far it became a fashion statement. And yes, this was his deliberate strategy. I like to imagine that back home in Zürich he stands out against the more formal style of the Swiss by slouching around in jogging pants, but in reality I don't think he goes beyond removing his tie and undoing a top button on his shirt.

Helbing's vision for big data is as much social as technological. Seeing that it's only too easy to collect information on our every movement, communication and decision, he fears a future in which governments and large corporations exercise surveillance on a scale undreamed of in the most dystopian science fiction.

He received a speeding fine for exceeding the speed limit by 1.6kph (1mph), the transgression captured and the fine issued automatically, by machines. This, for him, is an example of how excessive regulation has a corrosive effect on society. If we cannot take risks or make mistakes, how will we continue to progress? Even seemingly benign initiatives designed to nudge us in the desired direction can erode our capacity to weigh up conflicting ideas and make good democratic decisions.

'We are entering a world that will be ruled by a different logic,' he writes in a 2014 article entitled 'Genie out of the Bottle'. 'But who will rule it? Big business, governments empowered by Big Data, a computer-based artificial intelligence, or its citizens?'

But Helbing also sees the potential social benefits of sharing information to solve collective problems, like the aftermath of an earthquake. Instead of a top-down model, in which all our data is fed upwards to a central brain, he proposes a decentralised grid, a citizen web, in which we all control our own data: 'An era of "collective intelligence" ... an age of creativity, participation and well-being.'

One of Helbing's main concerns is trust. If we don't trust governments, society breaks down. Since governments can so easily amass data about each one of us, it's important that we know how they are collecting it, using it and with whom they're sharing it. So he argues that we need more transparency, more accountability to the public. With open data must come more openness about the human decisions involved in collecting, choosing and using that data.

Living in glass houses

Who could object to more transparency? After all, our elected representatives, and our paid public servants, are working in our name, spending our tax money, to shape the future of the society in which we live. Without knowing what they're up to, how can we claim to be in democratic control?

'My Administration is committed to creating an unprecedented level of openness in Government,' US President Barack Obama wrote in a memo published on the internet for anyone to read.[*] 'We will work together to ensure the public trust and establish a system of transparency, public participation, and collaboration. Openness will strengthen our democracy and promote efficiency and effectiveness in Government.'

He goes on to describe a vision that echoes Dirk Helbing's ideas, all about collective expertise, opportunities to participate in policy-making, and accountability to citizens. And the Electronic Frontier Foundation (EFF) agrees that transparency is important – especially: 'given the government's increasingly secretive use of new technologies for law enforcement and national security purposes.'

If you're watching us, we're watching you.

I certainly support greater symmetry of information. As I write, the UK is discovering that our secret service has routinely been intercepting all our telephone communications for the last 10 years. This wasn't technically illegal, but it certainly wasn't debated in parliament before being authorised.

I don't want to be killed by terrorists, but I do want those who claim to protect me to be accountable. Otherwise, the democracy they're supposedly protecting starts to wear a bit thin. So laws such as the Freedom of Information Act, giving me, or anyone, a statutory right to get answers from public bodies, are a good thing, in my view.

Open Data goes one stage further, publishing information in an accessible form before anyone has to ask.

Transparency is almost universally acknowledged as an obvious good, like sunlight. 'Sunlight is the best disinfectant,' said American Louis Brandeis over a century ago, referring to corruption in banking. Like sunlight, transparency exposes

---

[*] https://www.whitehouse.gov/the_press_office/TransparencyandOpen Government

corruption, laziness and incompetence to public scrutiny. But there is more to accountability than publishing all your data.

I've had a number of very informative off-the-record conversations for this book, with people who were keen to tell me what they knew but couldn't go public as the source of my information. There's no digital data on these conversations. I didn't put the names, dates or locations into any kind of electronic diary or calendar. I didn't even record them on an old-fashioned – analogue* – magnetic recording tape. I scribbled some words in a notebook with a pen, like a journalist from the previous century.

If the entire process of this book were transparent, I couldn't have done that. Those people couldn't have spoken to me without potentially risking their jobs, or at least giving themselves a tough time at work. So that's one problem with assuming that transparency is always a good thing. But it's not the only problem.

One of my conversations was with somebody who works in the financial sector. Since the financial crash of 2008, the sector is highly regulated, and my source's employer had to send all their transaction data to the national regulatory body, to check that they were complying with all the regulations. Not just one, but several copies of the whole massive dataset: amounts, dates, times, everything. All the data you could possibly need to spot irregularities and maybe head off a future financial crisis.

After a few months, one of the other employees looked at the data they had been sending to the regulator, and found that they had apparently not been complying with the regulations. In fact, it looked as if they'd been non-complying in a big way. Which was worrying for a large financial institution.

---

* If you're idly wondering about the difference between analogue and digital, it goes back to the ones and zeroes. Digital recordings use the '1 or 0' language that computers can handle easily. Analogue recordings use wave signals, sound waves transformed directly into varying waves of electrical activity. Analogue uses continuous waves, digital uses discrete bits of data.

Careful checking revealed that the problem lay not in what they'd been doing, but in the way the data had been recorded and processed. Columns had been swapped, in effect, showing the opposite to the true picture. They were complying with the rules after all.

However, the supposed regulator, whose job was to keep a stern eye on irresponsible financial companies, had not said a word. They hadn't even noticed that the data coming in seemed to show serious rule-breaking. Which is worrying, but not entirely surprising when you consider the volume of information pouring in every day from every single company.

If you want to find more needles, bigger haystacks are not necessarily the answer.

The financial regulator may have relied too much on knowing that all the data was being piped straight into its electronic vaults, with the attitude that nothing is being hidden, so everything must be OK. In the same way, transparency may be mistaken for accountability.

Do you trust me?

I'm going to ask you some rather personal questions now. But it's OK, they're rhetorical ones, you only have to answer them in the privacy of your own mind. At the time of writing this book, nobody has technology that can read your thoughts, so you can be completely truthful with me, knowing that I'm just a literary device in your mind.[*]

---

[*] Not the 'I' who is writing these words, I am a real person. There's a plastic keyboard here that makes noises as I type, and has old bits of fluff caught between the letters, and I'm sitting at a desk that physically exists. Otherwise this whole book would be a figment of your imagination. Which would be impressive. And would have saved me a lot of work. But the 'I' to whom you address your answers is in your imagination. Unless you decide to send them to me via social media, which I don't encourage.

Are you married, or in a similar close relationship?

Do you own a smartphone? Is it always, or nearly always, with you?

Would you be happy for your partner to be able to track your smartphone's location at any time? Would you want to track their smartphone?

If you answered the last pair of questions with, 'What? No! Why would anyone do that?' then I'm with you. That crosses over my line from sharing to stalking.

But it seems many other people will be thinking, 'Can you do that? That would be awesome! Hey, honey, did you know we can track each other's smartphones at all hours of the day and night!' Or even, 'Yes, of course. We already do that. And all our children too. And the cat.'

I was on a train back to London with a friend, enjoying a cheeky beer after our hard day's work, and discussing our plans for the evening. Wondering if she had time to meet her husband, she got out her smartphone. But instead of calling him, or sending a message, she simply checked his location.

I was shocked. Bemused by my response, she told me that several of our mutual friends did the same with their husbands and wives. To her, it was a matter of convenience, like having a shared diary to plan social events. 'But we trust each other,' was her attitude, 'why wouldn't we give each other access to our phones' location?'

My attitude, by contrast, was that trusting somebody was exactly the reason you might give, and expect, some privacy. Somebody who wanted constant access to every moment of my life would feel, to me, like somebody who didn't trust me.

Nor do I want to know everything my loved ones get up to. And not just because it would spoil the surprise when I open my Christmas presents. I like to leave them space to be themselves in ways I might not want to witness.

I'm not casting any aspersions on my friends' marriages. I see no evidence of suspicion, distrust or any reason to distrust.

There's nothing sinister in what they choose to do. But it did provoke some thoughts on trust and transparency.

In his satirical science fiction novel, *The Circle*, Dave Eggers describes a world in which sharing everything by social media becomes so much the norm that politicians start to 'go transparent'. They announce that they will wear cameras that stream everything they say and do to the internet in real time. The people they represent would no longer need to scrutinise their past record on a website, they could instead be witness to every conversation with campaigners, researchers, fellow politicians, as they happen.

You may be thinking you'd like that. No more off-the-record conversations behind closed doors with shady characters who lobby on behalf of tobacco companies, arms dealers, or animal rights groups. No more unwritten deals with trades unions for their members' political support, or with wealthy donors. No more nasty shocks when the newspapers reveal your friendly local candidate gets drunk and uses racy, or racist, language in private conversation.

No more private conversation, in fact.

So, for a start, no off-the-record conversations like the ones I have, where people tell me frankly what they know, or what is on their mind. Because transparency for me means transparency for everyone I speak with. Want to tell your MP about the terrible working conditions in your warehouse, but don't want to lose your job? Bad luck, your conversation will be live-streamed to the internet.

Also, no frank conversations with colleagues about how to campaign. Politics is adversarial, it's a clash of interests and ideas, and you may not always want your potential opponents to know every stage of your thinking. Not because you plan to lie, or to take bribes, but because you have to decide your strategy and tactics.

Most important of all, no free debates about ideas.

Say somebody in your party has a new, radical, some might say bonkers, idea about how to reform welfare payments. Do you want to have that aired in public, before you've all had a chance to say, 'Whoa! Dave, that would never work! Here's why …' or 'That is against every principle our party stands for!' or possibly, 'That's a crazy idea. But it has made me think of a new approach which might just have something …'?

Or, worse still, do you want that person to think it but never say it, because they fear the public response? Better, surely, to discuss it in private, work out whether it's a terrible idea, or the germ of a brilliant one, or maybe both.

If everybody is perpetually watching what they say for fear of public exposure, that kind of honest hammering out of new ideas would be impossible. Instead, we'd get more and more of the same, pre-packaged before it left anybody's mouth for maximum public acceptability. Which feels a lot like the place towards which we're already heading.

I'm not convinced that making everything transparent will necessarily foster more trust in our politicians. Calls for more transparency seem to me more like a symptom of lack of trust.

But what about those policies, ideas and manifestos? Can big data help us, the electorate, make an informed choice in the polling booth?

Show us your evidence

As the 2015 general election approached, I went along to a meeting organised by the Alliance for Useful Evidence, hoping for some answers. First to speak was Ed Humpherson of the UK Statistics Authority (UKSA), who started by waving a book at us, *Great Speeches of the Twentieth Century*. The inspiring words of people such as JFK, Martin Luther King and Winston Churchill made almost no use of numbers, he pointed out. Today's speeches, by contrast, are very focused on quantification, statistics and facts. 'Drowning in facts' was the phrase he used.

The statutory role of UKSA is to 'promote and safeguard … statistics that serve the public good,' though not, as Ed pointed out, to police political discourse. And he identified some areas where official statistics are misused or misinterpreted. But when he ended with a joke that perhaps the equivalent book waved around in 2115 would be *Great Fact Checks of the 21st Century*, I think he too felt that would be a loss.

The Alliance launched their Manifesto Check at the event: 'an unbiased evaluation of the policies as they appear in the party manifestos', and a service enlisting policy experts to check facts suggested by members of the public.

It also promised me 'evidence-based comment and analysis', and 'evidence-based discussion'. I do want to be able to call on evidence, or statistics, or data, to assess how successful policies have been, and compare claims about future policies. If I'm not sure which party to believe on how many new houses we need to build, or how much they will cost, I can refer to some experts who have given the figures a going-over more thorough than I have the time, or statistical skill, or specialist knowledge, to do myself. But I'm not sure about 'evidence-based discussion'.

It seems pretty common sense to want political debate to be based on evidence. If somebody claims that reoffending will fall if all prisoners are electronically tagged on release, and followed by a drone that films every second of the rest of their lives, or alternatively by making sure that every prisoner has a secure home to go to on release, I'd ask whether these claims were based on anything that has been tried in the real world.

However, even if you showed me studies that found a 90 per cent fall in recidivism under the drone scheme, and only a 15 per cent fall under the housing scheme, I'd still vote for the help with housing. Not because my maths is rubbish, but because on principle I think everyone should be able to move on after serving their sentence, make a fresh start and enjoy a private life like everyone else. And also that ex-prisoners being homeless helps neither them nor the rest of us, regardless of whether they go on to reoffend later.

Another of the speakers at the same event, Director of the Nuffield Foundation Sharon Witherspoon, caused a little ripple in the roomful of statisticians and evidence-checkers. She stated that she doesn't personally believe in evidence-based policy. Evidence for policy, yes, and evidence-informed policy, but starting from an enlightenment point of view, 'the facts don't speak for themselves'. And I agree with her.

Data: fuel for democracy, not a satnav

You may be thinking that adversarial politics is a bad thing, and that it would be much better if everything could be out in the open and we all worked together to find the best solution, with the aid of transparency, open data, and evidence-based policy. But all these things depend on a few assumptions.

First, they assume that data is always more objective and neutral than opinion or argument. If we start from the evidence, the data, at least we can all agree on that. It's made of numbers, so it must be true and free from human bias.

This is part of the appeal of big data. Because it is generally gathered by machines, it's seen as being less fallible and less biased. It's too easy to forget that some human beings came up with the idea of gathering this data, using this method, and putting it into this form. Why ask one question and not another? Why measure this dimension and not that one? There are probably good reasons, but they all represent human decisions.

Second, they assume that data about how the world is now, and how it was in the past, contains the answers to the big questions of politics and policy: Why did these things happen? What does the future hold? What choices can we make, and what are their likely consequences?

In reality, statistical data can describe many aspects of the world well, with precision* that may be misleading. But

---

* There's a useful distinction between precision and accuracy. If I tell you that I am typing this chapter at GPS coordinates 40.7484° N, 73.9857° W, I am giving a very precise position. It's the Empire State Building in New York. Precise, but not accurate. I am really typing in London, a description of a location that is less precise but much more accurate.

explaining the world is altogether different. Why is crime falling? Nobody really knows. How can we make it fall still further? Hard to say without knowing why it's already falling. What will happen in the future? Again, very hard to say with confidence, because without a causal explanation, all we can do is assume that current trends will probably continue.

That works well in the natural sciences, where a falling apple continues to fall until it hits either the ground or Isaac Newton. In the human world, people tend not to do what they have always done, luckily for me. If they had, I would still be living in a cave, without being able to read or write, let alone type this chapter.* I would never even think to scratch 57 tallies into a wolf bone, if nobody had done it before.

Third, they assume that there is one 'best solution' on which we would all agree, given enough evidence. This rarely happens. Solutions that are in your interest may not be in my interest. Depending on your circumstances, your political priorities may be very different from mine. As somebody too lazy to have children myself, I might not want the government to allocate all available cash to paying for childcare and a free beer allowance for all parents.

Even if we have the same circumstances, our ideas of what constitutes a 'best solution' may diverge. As it happens, I am profoundly grateful to parents for raising the generation who will be programming the robot to look after me in my decrepitude, so I don't begrudge them extra resources. I'd happily vote a beer allowance for all new parents. My material circumstances may affect my opinions and attitudes, but they certainly don't determine them.

You may disagree. What then happens, in a democracy, is that we both try to win a majority of people to our side of the argument, which is why politics is adversarial.

---

* So I would be a mere figment of your imagination. Except that you'd also be living in a cave, so imagining a woman sitting in London typing a book about big data on a computer would be quite a feat.

The very breakdown of traditional political loyalties that drives politicians to relate to us through data is a tremendous opportunity.

We could step away from the relics of twentieth-century politics and make a fresh start. We could talk about the principles we think are important, the different visions we have of how the world might be better, and start some real debate. We could use the conflict between our ideas to test what we think, refine the good ideas and throw out the bad. We could seize some of the opportunities that technology gives us to transform our world, with bold visions and a spirit of experiment and risk-taking.

We have to decide which we want:

'We do these things ... because they are hard' or 'We do these things because they are evidence based.'

'I have some data ...' or 'I have a dream ...'

PART 3: BIG IDEAS?

Whoever scratched those notches in an Ice Age bone could never have dreamed of where that idea would take the human race. From a simple tally to machines that collect information, process it and predict the future with scarcely the touch of a living hand, it's been quite a ride.

And it's only just beginning. By the time you read these words, even the extra 'update' chapter will seem outdated, as what I said might one day be possible becomes the everyday. And something I told you was impossible has already been achieved in some lab in California, no doubt.

Big data seems to hit the news every day, from fears of domestic appliances like washing machines and sex toys being hacked, to researchers using it to analyse the evolution of jazz music.

So I can't possibly tell you everything that big data will do for us, or even everything it has already done. But that's fine, because you can all read, so you can follow the news for yourselves.

I wrote this book, not because I want to turn you all into data analysts,* but because I think big data throws up some important questions that we should all be talking about. So in the last three chapters I want throw some of them at you.

After spending a few years thinking about this topic, and reading about it, and having conversations both public and private, I have formed some opinions of my own. I don't

---

* Which is just as well, because I can't write computer code either.

expect you to share them all. I'd much rather you treat them with as much scepticism as everything else in this book.

My hope is that they provoke you enough to go out, start a conversation, and forge your own ideas about big data, and what it could mean for our future.

CHAPTER EIGHT

Big Brother?

From the centre of San Francisco's Downtown, where locals and tourists fill the sidewalks below the tall buildings of the Financial District, I descend to the BART station and board the underground train. A few minutes later, we emerge into a horizontal landscape, where shipping containers stretch away in all directions. The only things against the blue sky are white-painted cranes, which seem to gaze back across the Bay towards San Francisco like lonely metal horses. Rail tracks and roads cross the barren concrete, but hardly a single human face peoples the sprawling port.

Most of the massive container ships that pass under San Francisco's Golden Gate Bridge are heading to or from the Port of Oakland. It covers 32km (20 miles) of shoreline, and more than 2 million containers a year pass through Oakland between rail, road and ship. My train passes acre after acre of multicoloured boxes.

Finally, we start to pass signs of habitation: low-rise houses, cheaply built, with old vehicles parked between them. Painted in large, angry letters on the side of one building is the slogan, 'Black Lives Matter'. Then it's back into darkness.

Shortly afterwards, I walk out of Oakland City Center station into a cosy enclave of cafes with tables outside, nondescript shops and empty pavements. Many of the shops are vacant, and the fountains are dry, but it doesn't feel like one of California's most crime-ridden cities. I could be in any small town in England.

Panopticon

Across the road, in Frank Ogawa Grand Plaza, I meet Brian Hofer on the lawn in front of City Hall. He's a local attorney, and he's here to tell me the story of the Oakland DAC.

'The Domain Awareness Center, or the DAC, was a port infrastructure improvement project,' he says. 'At some point the project expanded to become a joint project of the city and the port that would include facial-recognition software, automatic licence-plate readers, ShotSpotter, 700 surveillance cameras throughout Oakland unified school district, Oakland housing authority, 300 TB of data storage, along with other benign things like vessel tracking, tsunami warning, earthquake warning.'

Like a more ambitious version of Glasgow's smart city, but with less emphasis on bins and heating, and more on gunfire and earthquakes. And I have heard that Oakland, however calm it appears, has a gun crime rate among the highest in California.

Brian is sceptical. 'Supposedly we're up there, yes. Unfortunately, part of those statistics are based on ShotSpotter, and it is one of the most inaccurate pieces of equipment ever invented. The most recent audit I saw had a 7 per cent accuracy rate.'

What is ShotSpotter? 'It is an acoustic sensor designed to identify gunshots. The need for this according to ShotSpotter, and it is true in Oakland, the community often does not report gunfire to Oakland Police Department.'

Why not? 'Many reasons. One: they don't show up, so why call? Number two: there's this distrust in certain communities, certainly communities of colour that have been brutalised in the past and somewhat today, that they don't want to call the cops.'

The idea is that ShotSpotter can locate the gunshot and alert the police directly, so they can arrive in time to arrest the shooter. But, says Brian:

'It doesn't work that way. Not only does it not identify gunshots accurately, it does not get information to the police on time. The officers don't show up, when they do show up there's no one there. It's not a good piece of equipment. I think it's more of a taxpayer boondoggle than anything else.'

'But we do have evidence here in Oakland that under the right circumstances it can record a human voice. Oakland Police Department used that favourably at a trial where someone was shot. On the ShotSpotter audio you could hear him identify his killer, the shooter, and that person was arrested and prosecuted successfully. So there are success stories but I get back to my 7 per cent accuracy rate: it's just not worth the money.'

And it's a system that can record people's voices as they walk around the city?

'Under the right circumstances. I'm not comfortable saying that it could always be on. I do think if the software was reconfigured that is a possibility, yes, you have a citywide surveillance network of microphones. I think there is a certain threshold, a certain frequency needed to trigger it right now. But without an independent audit, it's hard to confirm that.'

Community distrust of the police is not unfounded. On New Year's Day 2009, a 22-year-old passenger called Oscar Grant was pulled off a train at Fruitvale Station by a BART police officer, and shot dead while handcuffed and face down on the platform. Grant was African-American. His killer, who claimed the shooting was an accident, was convicted of involuntary manslaughter.

In October 2011, a group of protestors set up an encampment where we are standing, in Frank Ogawa Plaza, renaming it Oscar Grant Plaza. Calling themselves Occupy Oakland, they aligned themselves with other Occupy protests around the US and beyond. City and police efforts to remove the encampment escalated into violence and exacerbated tensions between authorities and the community.

Into this context the city introduced a proposal to link up existing surveillance equipment with new technology, all feeding one central Domain Awareness Center.

'It had originally been sold as a port infrastructure project, then it was sold to us as this thing for first responders to help with efficiency. The problem being, the only time our previous version of this, the Emergency Operations Center, had been activated was in response to protests. So we had some suspicions.'

As Brian walks me through quiet streets towards the DAC, he talks about the city's motives.

'Because there's distrust among a lot of the citizens with reporting crimes or being witnesses, Oakland has made this decision that they're going to use technology. It's shiny gadget syndrome. We're up the road from Silicon Valley, everybody's promising us all these wonderful things, this big data-driven solution that is going to solve all our society's problems.'

So that's the push. Then, says Brian, there's a financial pull. 'The smartest thing – this is me putting my little conspiracy hat on – is that the Department of Homeland Security is funding this via grants. So when I walk before City Council and try to make a taxpayer argument it's really hard to show a local pain point.'

'Where the taxpayer harm does come,' he says, 'is all those ongoing costs: maintenance, software, the staff, that we end up spending in the future after that grant money is gone.

That's what we're trying to show: it's not cost-effective, it's not doing the job, and it's impacting my privacy.'

A few blocks from City Hall, we reach the building housing the DAC.

I feel slightly disappointed. Having read about its planned capabilities, I'd pictured something between the Pentagon and a Bond villain's lair. Preferably an impregnable steel fortress inside an artificial volcano crater, but at the very least a sheer glass tower guarded by enigmatic men in sunglasses.

This is a low-rise, nondescript, white-painted building with absolutely no security around it. In fact, it's a fire station.

'It's a working fire station. You can see on the ground floor there's a couple of fire trucks. I have a couple of buddies that work inside,' says Brian, 'and then upstairs there's the Emergency Operations Center. If there's a natural disaster, some kind of emergency, this is where the crisis team assembles.'

I ring the buzzer and ask if we can come in and see the EOC, but the person inside says, apologetically, that I'd need to make an appointment. Brian tells me there's not much to see anyway.

'You see it on television shows, crime shows, it's just like any other little command room. It's a row of chairs with desktop computers and then one big flatscreen TV for everybody to look at.

The DAC is just an application, it's software that sits inside these computers that has way more capabilities at aggregating all the data inputs,' he explains.

'The EOC could see and analyse all this data but it's all on separate, distinct systems. What the DAC is doing is bringing it together all on to one screen and letting you overlay data so you can see a bigger picture. And of course that's wonderful in an earthquake scenario or a fire. You're able to track wind patterns, so if there was a chemical fire you could see which way the fire …'

… is going to go before it happens. And send fire trucks there, and save lives?

'That's wonderful, that's great, we have no problem with that. No one has ever made an argument before City Council against those types of warning systems.'

What is the problem, then, if it's just a system for dealing with earthquakes and fires?

'That ability to aggregate all the data also allows you to create a mosaic, to see the patterns in someone's daily travel habits or their life. Oh, Brian is going down the marijuana dispensary, now he's hanging out at the abortion clinic, or he's at that Occupy protest because we're also tracking licence-plate numbers. And so the good is also the risk in this type of capability,' he says.

'Of course we had a big 1989 earthquake, freeways fell down, radios weren't working, it was a bit crazy here. So to be able to coordinate and move resources around faster would be wonderful. But with that you need to have safeguards built in, to address the civil liberties concerns.'

Even the DAC'S location in a fire station set Brian's antennae twitching.

'It's an interesting trend that we've noticed. A lot of the DAC's funding is coming through fire departments and not police. We're not quite sure why. I read a somewhat alarming White Paper out of the Monterey Naval Academy think tank, about using fire departments to get around the Fourth Amendment.'

As a Brit, I have to ask what the Fourth Amendment is.

'The Fourth Amendment is protection against general searches and seizures. That's where the requirement to get a warrant comes from, basically because of the British. Judges or magistrates were authorising these general warrants so you could go search anything at any time. That was one of the causes of the American Revolution, and the Americans pushing back against the Brits. So it's kind of a big deal.'

So in a way, we helped America build freedom into its constitution?

'Anyway, this Naval Academy White Paper was saying that you could use fire departments to get around the Fourth Amendment. You know, maybe it's a marijuana grower's house or something, you see some evidence of wrongdoing, you didn't need a warrant to get in there, and then they could report back to the police. So we have some concerns over data-sharing between different entities.'

Americans have a number of legal principles from which to defend their privacy. As well as the Fourth Amendment, which demands a specific reason and target for a search warrant, they can invoke the First Amendment, which protects freedom of speech and association. California is among the states that also enshrines the right to privacy in the state constitution.

Privacy and free speech are sometimes discussed as though they're conflicting principles. Brian would disagree:

'I went down to Cuba in 2008. It was slightly opening up, people were very willing to talk to us, but they wouldn't say Fidel or Raoul's name in public. They would say the One or the Two, they'd make a gesture, the beard … The younger guys, out in the middle of a baseball field where there are no microphones, were like: There are tons of informants.'

In every town there were community organisations, and those were the informants, the locals told Brian. 'These people will inform on everyone, and that's why no one's ever challenged the Castros' power. They were masters at keeping the community disjointed so there was never any opposition.

And that's what it's about, mass surveillance. Of course there is legitimate criminal surveillance, like bringing down the mafia, but dragnet surveillance is about population control, and it's very effective.'

Nor is Brian convinced by the argument that those who have nothing to hide have nothing to fear.

'We do have something to hide. Not just our bank pin numbers, but same-sex relationships, marijuana use … America went crazy with marijuana prohibition even though everyone smoked. Everyone used it, but we had to pretend in public that we didn't.' And eventually, private use led to changes in the law.

'And same-sex marriage: you had to keep your relationship in hiding, in privacy. Because you had privacy, you were able to do that.

But without privacy, we would not be where we are today. You don't have that ability to form a different opinion, you have to be completely homogenous, like every other person.'

Even without laws explicitly forbidding certain actions, words and ideas, constant scrutiny has a chilling effect. That's why freedom of association depends on the right to privacy, and freedom from surveillance.

'I thought we rejected that in America, but it's creeping back this other way.

There's a famous Supreme Court case of people trying to get the NAACP* to reveal its membership rolls, so they could target these people. And the Supreme Court said: No, they have freedom of association,' he says.

'And then, from the other end of the political spectrum, here in California with our Proposition Eight fight over same-sex marriage, the lefties, the progressives, were trying to force the conservatives to reveal their donors and membership rolls. And we're like: No, we already decided this issue!

It's like we didn't learn much. Whether you're left or right we're still going after this freedom of conscience and trying to get rid of it.'

And Brian warns that the ease of aggregating information through technology poses a new danger.

'Nowadays, sure, the NAACP doesn't have to turn over their membership roll but the NSA† can get in their computer anyway. Or they just use a Stingray and intercept phone communications while sitting outside your building. They use a licence-plate reader and drive around the building and look at the licence-plate numbers. So effectively court decisions are meaningless if surveillance equipment is used indiscriminately.'

Stingray is a word I keep hearing. I ask Brian what it means.

'Stingray is the brand name of an IMSI catcher. If you take off the back of your cellphone and lift up the battery you'll see this little IMSI number, we all have these unique identifier numbers in our cellphones,' he explains. 'Using a Stingray you can identify this unique number and by driving around in ever smaller circles you can triangulate someone's position. So if you've got a fugitive on the run, kidnapping victim, maybe

---

* National Association for the Advancement of Colored People, a leading force in the struggle for equal civil rights for all races in America.
† National Security Agency.

you've got an old person suffering from dementia that's wandered off, we now all have mobile tracking devices on us: our cellphones, we all carry them.'

Although old people can be frustratingly unwilling to carry their cellphones everywhere. Perhaps they know something we don't? Who knows what they're getting up to.

Brian continues explaining how Stingray works:

'A weakness of all cellphones is they're constantly looking for the strongest tower, that's how they connect, so the Stingray sends out this massively powerful signal and forces the phones in range to connect to it. Even if they get a warrant for my phone number it's still a general search because it has to intercept all of your data and all the other phones that are in that range.'

And if that's not intrusive enough, says Brian, 'In certain configurations it intercepts metadata, the phone number, the duration of your phone call, but nowadays – Oakland police department recently acquired this capability – it can intercept content.'

Which sounds a lot like the kind of general search prohibited by the Fourth Amendment.

'By federal law, when Stingrays were sold to local police departments they were not supposed to have this capability. You have intimate photos on your phone, you have private messages that can now be intercepted if you're just walking randomly down the street,' says Brian.

'You wouldn't even know you'd been intercepted, that your phone is being searched, because it's an invisible signal and you just happened to be walking in range. You're not even making a call. Your phone could be off and it can still be intercepted. It can intercept a text message, voice content.'

If you're now thinking you don't mind, because you have nothing to hide from the police, Brian would like you to think again.

'It's not just an activist thing, think of any doctor-patient conversation you've ever been involved in, whether it's yourself or your kid. It is obviously private, you don't want your medical information being intercepted. Attorney–client communications. Stingray signals are so powerful they penetrate the walls. All these attorneys sitting inside, using their cellphones for privileged communication?' Brian gestures at the office buildings around us.

'It's at risk of being intercepted. There's hardly anything more sacred under American or just Western law than attorney–client or doctor–patient privilege and all that is at risk now. This isn't just about occupiers and activists, it's an invasion of everyone's privacy and dragnet is the only way this technology can operate.'

I start to wonder whether technology, which has been the villain in most of this conversation, could help protect privacy. Could you configure a Stingray so the software would dump all the data except the number for which you have a warrant, before a human being could see it?

'Right. That's where I think we're going to get. Because Stingrays are highly effective. They always find the phone, they always get their guy. So I don't think the majority of citizens would reject Stingrays outright. Oakland Police Department has told me that they immediately delete all the other data – we want to independently audit that of course and make sure there's penalties if that's abused but ultimately I think that's probably where we'll end up.'

You may have guessed by now that Brian, and other Oakland citizens, did not welcome their city's data-driven surveillance network. Later in this chapter, I'll come back and finish the DAC story.

The spy who bugged me

Oakland's resistance to being watched over was given momentum by revelations in 2013 about how much data government agencies were gathering on their own citizens. These revelations came from a former CIA employee called Edward Snowden who turned whistleblower in 2012.

The inside information he revealed to journalists is still emerging.

Snowden himself is, as I write, living in exile in Moscow. In case all the fuss passed you by, here are the basics of what he made public.

Shortly after the Second World War, the UK and US set up a spy network called Five Eyes with their allies Canada, Australia and New Zealand. Still going strong, it now uses technology to intercept and monitor communications between people all over the world. The NSA's Prism and GCHQ's Tempora are part of Mastering the Internet, which collects information in two main ways.

'Upstream', as one of the NSA's leaked files calls it, means tapping directly into the hardware that carries your communications, the modern equivalent to putting clips on telephone wires to listen to your calls. Today, they're more likely to be intercepting signals passing through a fibre optic cable under the sea, or planting software in your computer that reports directly to them.

Prism goes straight to the companies who already have your data collected and stored. Companies like Google, Apple, Microsoft and Skype who provide your telephone and internet services. Prism doesn't just go back through old records, it can register in real time that you're making a call or interacting on a chat forum.

A Pew survey in 2015 found that 87 per cent of Americans had heard about government surveillance of telephone and internet communications. Of that 87 per cent, a third had done something about their own privacy, mostly simple things like better passwords for email accounts, changing the way they search online, or having more conversations face to face.

Two-thirds of those surveyed said they were losing confidence that the surveillance programs serve the public interest. But they tended to support surveillance of suspected terrorists, and to say they weren't personally concerned about their own communications being monitored.*

The British public are similarly ambivalent about government eavesdropping, and not because they're unaware of how much GCHQ can collect. If anything, they may overestimate the capacity, or desire, for such mass surveillance. In 2015, almost two-thirds said yes to the YouGov question, 'Do you believe that GCHQ has the resources and technical capacity to intercept/collect the internet-based communications of every British citizen?'

Naturally, neither GCHQ nor the NSA are going to reveal exactly what they can collect, or how, or whether they do. Intelligence services have complained that Snowden's revelations drove terrorists and other targets to use more sophisticated techniques.

Government prying into its citizens' communications is a genuinely tricky question. I feel very strongly that I don't want anyone to access my private life without my express permission. But I also feel strongly that I don't want to be killed by terrorists, and quite strongly that I don't want to be the victim of a major crime.

In *Orwell versus the Terrorists*, British researcher Jamie Bartlett writes: 'We demand perfect security, but thanks to the Snowden Effect, that's going to be harder to achieve ...'

He cites a growing consensus against mass surveillance, increased awareness of technology to elude it, and a shift in terrorists' tactics:

> *And yet simultaneously, we have an impression that the security services can see everything, and so should stop everything, which is*

---

* Then again, if you were concerned about your own communication being monitored, you might not say so in an online survey. Wouldn't that make you look as if you had something to hide, and draw unwanted attention to yourself?

impossible. The Snowden revelations have created a false impression
that the intelligence agencies are monitoring every single thing we
do online, our every click, swipe and movement. And the resulting
opinion shift against internet surveillance limits the space the
intelligence agencies can operate within.

And because of the nature of online data – the fact there's so
much of it out there – there will always be some clue, some digital
breadcrumb, that's missed. More data doesn't always mean more
insight: it can also increase 'noise', making the 'signal' harder to
pick out.

Bartlett gives the example of Michael Adebowale, who
expressed via Facebook a desire to murder a British soldier,
and six months later did murder British soldier Lee Rigby. If
GCHQ is monitoring everything, many asked, how could
they miss that?

If we continue as we are, warns Bartlett: 'the result will
be an intelligence agency that is seen as both omnipresent
and incompetent, one that lacks broad public support
and can't do its job. This is the precise opposite of what
we want.'

Instead, he calls for a shift away from mass surveillance by
computer towards human intelligence that targets individuals,
and for clearer public oversight of the powers exercised by
intelligence agencies.

Outlaws and back doors

It's easy, with hindsight, to ask why nobody picked up a post
from a future murderer. But most people who express the
desire to kill somebody don't go on to kill somebody.

When the UK was hit by heavy snow in 2010, Paul
Chambers tweeted: 'Crap! Robin Hood* Airport is closed.

---

* Yes, the UK has an airport named after legendary outlaw Robin Hood.
Which gives this story an ironic twist.

You've got a week and a bit to get your shit together, otherwise I'm blowing the airport sky high!' An over-emotional reaction to a potentially delayed flight, but he was travelling to meet somebody special. Two years and two appeals later, Paul was finally acquitted of 'sending a message of a menacing character' as the High Court judge accepted it was clearly a joke, sent using Chambers's real name and not directly to airport staff.

When a 14-year-old in the Netherlands tweeted a mock bomb threat to American Airlines, they responded that they had forwarded her IP address* and details to the FBI. She handed herself in to local police in Rotterdam after tweeting, 'my parents are gonna kill me if I tell them this omg pls' and 'I need a lawyer. Any lawyers on here?'

It's easy to read these accounts and laugh at the idea of taking them as serious threats. But if social media accounts are monitored not by reasonable people, but by software, which is notoriously bad at irony, these people would be flagged up as potential dangers.

Though the UK legal system allows the police to arrest people for tweeting jokes, and the courts to pursue a case for two years, so perhaps the machines are not the problem.

Writing about big data and privacy feels like shooting at a moving target. When I started writing this chapter early in 2015, two UK Members of Parliament had just won a court case challenging a UK law called the Data Retention And Investigatory Powers Act (DRIPA), which gave sweeping powers to police and security services to get hold of our telecommunications data.

Where you were, when you sent emails, texts or telephone calls, and to whom, could be accessed on request, merely authorised by a colleague of the person who wants to spy

---

* Which identifies an internet connection, and hence often the user.

on you. This is often called metadata, because it doesn't include the content of calls and messages.

They would know I took photos late at night and to whose cellphone I sent them, but not what I was photographing in my bedroom at 2am. They would know which websites I looked at, but not exactly which page of OccupyOakland. org or icreacharound.xyz[*] I visited.

Even a mere human can infer a lot from metadata. A computer, combining many sources and looking for probabilities of ill intent, can infer plenty. Former head of the CIA and NSA, General Michael Hayden, is on record saying: 'We kill people based on metadata.'

As I return to this chapter a few months later, the UK government has published a draft IP Bill to replace DRIPA. Civil liberties groups are already pointing out how extensively police and security agencies would be able to access not only metadata, but all our communications, even encrypted messages, which the service providers would be compelled to decrypt on demand.

Encryption, which prevents anybody except the sender and the intended recipient from reading what is sent, has been an annoyance to some since Phil Brandenburger bought that Sting album online with his credit card.

Criminals would love to hack into your computer and get access to your online banking details, to your photographs in case there was anything blackmail-worthy, or just to your friends' email addresses, to tell them you've lost all your money in a distant country, and please can they wire you a few hundred dollars via this bank account?

Encryption makes it easy to lock up a file, and remarkably hard to unlock it. It's central to the relationships of trust that

---

[*] A site set up by Brett Lempereur in Liverpool, to show every website he visits, and give an example of what 'collecting metadata' means.

let us do so much business online, or via apps on our smartphones.

UK Prime Minister David Cameron, in a speech in early 2015, said: 'the question is, are we going to allow a means of communication which it simply isn't possible to read? My answer to that question is: No, we must not.' Our government wants all service providers to keep a digital back door so they can, if requested, let the police or intelligence services read our private correspondence.

Apple, one of the companies currently offering end-to-end encryption, has warned of 'dire consequences' if it was forced by law to end it: 'Any back door is a back door for everyone. Everybody wants to crack down on terrorists. Everybody wants to be secure. The question is how.'

And if would-be terrorists know all commercially encrypted messages can be decrypted, they'll simply switch to other channels and be harder to find.

Furthermore, if you accept the argument that no form of communication should be opaque to the security services, you are saying that there should be no truly private conversation.

The database state

Having your communication intercepted by spies is an extreme scenario. The same legal powers have been more widely used in the UK by the police against journalists, and even by local authorities checking that parents aren't trying to dodge school selection rules, for example.

That expansion of powers from anti-terror to anti-using-a-fake-address-to-get-into-a-better-school is more worrying to me than allowing the security services to snoop on our telephone calls. It suggests that anyone in authority feels able to check up on us, even for trivial reasons.

Different branches of government already hold a lot of data on each of us for legitimate purposes. Now there are calls to tie all this data together for the convenience of both citizens and state.

In some countries, all official records associated with one person are already linked together, with an identifying number. In Sweden, each resident has a unique number on the population register, kept by the Swedish Tax Agency. This includes your address and details such as the date you got married. All branches of government can use this register, including vehicle licensing, pensions and health care. It's also used by banks, insurance and mail-order companies, which must make moving house in Sweden a lot easier.

However, attempts to introduce a national ID card in Britain met resistance. Many people couldn't see the point of having another official document to prove your identity, when a passport, driving licence or credit card were generally accepted. Initial proposals to offer a voluntary ID card were dropped as potential costs grew out of proportion to perceived benefits.

In a London cafe I meet privacy campaigner Phil Booth, a key figure in the No2ID campaign against identity cards. Tall, with a dark ponytail, he talks with an intense energy and lots of swear words, some of which have survived on to these pages. He's not celebrating his victory against ID cards.

'I don't give a shit about cards,' he says. What he opposes is, 'the notion of the database state. That tendency of governments to try to run society or control people by watching or manipulating their data. That idea has been around amongst the bureaucrats since the invention of the filing cabinet: If only we had more data we could control things better!'

Phil answers his imaginary bureaocrats: 'If you are only engaged in collecting data in order to get into this big data thing, I suggest you should be thinking about what it is you want to achieve, and seeing if there is a smaller, more precise dataset, based in a good appreciation of the problem you're trying to address, that might suit your needs better.'

But surely, part of the promise of big data is recycling data you have already collected? If as a state you are already collecting people's tax records, driving licences and addresses, why not link it all together and save us all having to fill in four different forms? Phil's voice punches through the hubbub of the cafe.

'Because, number one: it is unlawful. Number two: it's an abuse of human rights.

This is why we have a human rights framework after the Second World War where a state went rogue and killed a lot of people. This is why we have laws in the area of data protection that recognise data that is about human beings, personal data, to be a special sort of data.

The rules governing personal data are very clear. If you're collecting personal data you have to do it for a specified purpose. That is fundamental. Collect it all, decide what to do with it later, is not acceptable. This is enshrined in the Data Protection Act.'

Phil would like to see a much clearer legal framework that links privacy, data protection and Human Rights. At the moment, laws governing data vary between countries, and are often ill equipped to cope with today's technology, let alone tomorrow's.

American law tends to focus on protection and redress from misuse of data, rather than regulating its collection. The European Union takes the opposite standpoint, but is currently rewriting its laws governing data, which vary between member states.

In Germany, for example, the 'right to informational self-determination' is already enshrined in the constitution. Personal data, concerning an 'identified or identifiable natural person' is covered by data protection laws, and special protection is given to certain categories of personal data. These include anything relating to somebody's racial or ethnic origin, political opinions, religious or philosophical beliefs, trades union membership, health or sex life.

Lots of different groups want the European Union to make better laws that resolve some of the issues big data throws up. Medical researchers want a new approach to consent, to make it easier to repurpose data collected for previous studies. Consumers want better protection against their data being sent abroad, beyond their control. And privacy campaigners like Phil want more control to remain in the hands of the subject of the data.

With fellow campaigner Guy Herbert he combined privacy and property in a new legal concept:

'Informational privity is analogous to some sort of human rights merged with a property right, where you can lease people use of aspects of your personal data. You may be able to say to someone: You can pass it on to other people. You can see how that plays out in health data.'

It's all right, I'm a doctor.

Health data is an area that brings the privacy dilemma into sharp focus. Neuroscientist Professor Paul Matthews talks about a fundamental dialectic, 'the two interests, first in preserving your privacy, but secondly a sense of altruism, a sense of public interest: in general you would like to help further medical research.'

Phil Booth's latest campaign, MedConfidential, was provoked by attempts to make all the medical data held by Britain's National Health Service (NHS) available to researchers for both health and policy research purposes. The first initiative, Care.Data, was put on hold when it became clear the public was not entirely willing to entrust their very personal information to unknown researchers for unknown purposes, at least without being convinced of the value and reassured about the risks.

'What I campaign on is not privacy,' says Phil, 'but confidentiality and consent. In the health arena, if you lose confidentiality, if you lose the ability for a patient to know that what they tell their doctor stays within those four walls, some people will withhold information that may be critical to their health and well-being, possibly even their life. And peripherally but equally importantly, to public health.

That is where privacy, data, the NHS, tips over into lives, deaths. People who think this data shit is not about life and death are sadly, woefully wrong.'

Phil can see the potential for research of using health data, but thinks that issues of trust were not taken seriously enough. One result is opt-out rates that threaten to make the data useless to researchers.

'The people who are in charge thought they could rely on the accumulated trust of generations in the NHS as an institution, to make use of the big data that an institution of that scale generates.

They failed to appreciate several fundamental points,' he says. 'One: this is not big data this is personal data. Number two: that trust cannot be assumed, in the same way that consent cannot be implied for anything other than direct medical care.

I would make a distinction between big data that is about non-human events and big data that is about human events. Human data is not the same because each piece of data relates to an actual person's life. Therefore it is, in law and in practice, in ethical and human rights frameworks, recognised as a different type and quality of data.'

'It is personal data if the person is directly identifiable, and it may be identifiable even if that data does not contain obvious identifiers like name and address, postcode.'

Phil spells out how easily seemingly anonymous data becomes personal: 'Because if you start to link together dated or timed episodes about human beings, that is a fingerprint about that person and that person could quite conceivably be re-identified. And in the context that might have other consequences.'

He concludes, 'These are not always bad consequences, but the people who are looking at this data need to have an ethical framework.'

Technology could help individuals exert more control over their own data.

In America, I chatted to Paul Terry, CEO of Vancouver-based company PHEMI, who had retired from tech when he joined a hospital board. Shocked by how poorly they used patient data, even in emergency situations, he got his team back together, 'like taking The Who back on tour,' he jokes.

PHEMI's system enables very specific pre-consent for research or medical use, and controls not only who can access it, but also when and where. So your consultant can read your file on their iPad over breakfast at home, but nobody who steals the iPad would be able to see your records. To Paul Terry, privacy is not about hiding your data from everybody, but about 'the right data to the right person, at the right time, in the right context.'

Phil Booth tells me about another company, Mydex, offering a Personal Data Store (PDS) that lets each person benefit from one home for their personal details, but stay in control of who else can see or use their data.

If the will is there, it's technologically possible for us all to retain some power over who sees our data, and how it is used.

Research design can also build in protection for the privacy of participants. Phil is impressed when I describe Eamonn Keogh's use of insects as roaming blood-samplers.

'That's a beautiful design to get around consent and anonymise at the same time. That's a perfect example. Well, I refuse consent –' Phil laughs and slaps his own arm – 'if I'm fast enough. But it's completely anonymous because that bug's belly is full of three people's blood.'

And we don't even know which three people they were.

Except, as Phil points out, if they analyse the DNA in the blood, it could be possible to link that sample to families or individuals whose DNA you already have on file. You know only that they were bitten by this insect within this time frame, but it could help focus the search for an elusive target.

Perhaps IARPA's funding of Premonition is not purely motivated by the altruistic desire to prevent disease outbreaks in remote parts of the world.

But sci-fi plots using airships, drones and mosquitoes are not what concerns Phil about big data and genetics. He worries more about 'starting to classify human beings in a certain way, by virtue of nothing more than a pure happenstance of their conception.'

Phil uses himself as an example: 'I would be in the genetic underclass, potentially, for some of the genes that happen to be in my family over which I have no control. Simply looking at a printout of what your baby's got the potential to have … These are not trivial decisions. We might have missed out on my aunt and cousin, myself maybe. My aunt is one of the oldest people alive with cystic fibrosis. At over 60 she is on her second pair of lungs, but she has lived a fantastically productive, wonderful, happy family life.'

I am reminded of the early statisticians, and those who first developed the science of genetics, and how many of their contemporaries embraced eugenics as the way to a better society. 'I develop that analogy with caution,' says Phil, 'but I'm genuinely watching this with very close interest at a political level.'

Given the ease and cheapness of sequencing human DNA, the genome genie is out of the bottle. Legal protection may be the only way to avoid discrimination.

America has the Genetic Information Nondiscrimination Act, GINA, which outlaws unfair treatment by employers or health insurers based on genetic tests. Designed for cases like women with a faulty BRCA1 gene, who might have been refused employment because of higher health insurance premiums, GINA was first invoked in 2015.

Two warehouse employees in Georgia were awarded $2.25 million damages, after their employer asked them for cheek swab DNA samples. The purpose was to identify who had been defecating in the grocery distribution warehouse. The two employees, neither of whom had DNA matching the offending deposits, successfully sued their employer for breaching GINA.

Not exactly the scenario envisioned when GINA was passed, but the principle is the same: just because the technology exists, you can't use it for any purpose you like.

Look me up sometime

When you meet an interesting* person, it's almost the default today to do an internet search for their name. The results depend on a few factors such as age and profession, but as well as career outlines and unflattering photographs, it's very easy to find private details such as siblings' names or previous addresses.

That's the dilemma of big data. To get the maximum use from it means linking up previously unconnected information, and analysing it in new ways. But that also means knowing more about an individual, more easily than ever before.

Most of us belong to different social groups: family, friends, workmates, people you went to school with and see once a year. With them, we express different parts of our personality. Do your parents know what you got up to at that party when you left school? Do your workmates know that you wear batman pyjamas?

Your answer to those questions depends, in part, on your age. My childhood, school and college years were passed in carefree anonymity, before social media came along to indelibly record my terrible fashion sense, drunken humiliations and awful, self-righteous ranting. If you're in

---

* I'm including both professionally interesting and romantically interesting in this, though personally I prefer to retain a little mystery if it's the latter. I try not to Google before a first date.

your twenties now, how do you quarantine the parts of your life whose main purpose is to help you grow up by giving you something to regret in future? Or do you just accept that everybody will know everything?

In his book, *Privacy,* Wolfgang Sofsky talks about the 'transparent subject', the person whose entire life is open to observation by the state. And this constant scrutiny is often welcomed by its subject, who wants protection, and is more afraid of fellow citizens than of the state. As Sofsky points out, the erosion of privacy goes much wider and deeper than the kind of all-seeing Big Brother described by George Orwell in his dystopian novel *1984.*

Rather than a powerful state insisting on access to our secrets, it often feels more as if we ourselves regard privacy as suspect, and insist on the disinfectant of transparency even for our own personal lives. Feelings must be expressed and discussed, family life opened to friends and experts alike, relationship status posted on the internet as a mark of commitment.

Do we not readily give away all sorts of information about our private lives when we post to Facebook, Twitter and the like? Our birthdays, family members, likes and dislikes, pictures of our homes, pets, friends, meals, weddings, babies, injuries, holidays …

As Sofsky puts it:

'Not the all-powerful "Big Brother" but rather many little brothers are busy finding out people's secret wishes and activities.'

Little brothers

As we've seen, predicting your customers' desires, sometimes before they know them, can give you an edge on your competitors.

Some of you may be more concerned about all the data gathered by companies to better understand you and market to you, than about what governments know. Others will feel that companies only gather what you agree to give them, and therefore anyone who doesn't like that should just stop registering on websites and filling in consumer surveys.

In practice, it's not a clear distinction. Hardly anyone reads all the privacy policies before clicking 'I Agree'. Once you've agreed that your data may be shared with trusted third parties, or not un-agreed by un-checking the box, you've relinquished all control over which charities, arts organisations or insurance companies can use that information to classify and then contact you. And in practice, whether through resignation or a feeling that we get benefits in exchange, most of us do share at least some data, voluntarily.

A Direct Marketing Association survey in May 2015 found that the commonest reason to happily share personal information was trust in the organisation asking for it. Given how much information most of us share, you may be surprised to learn that the least trusted organisations when it came to our data were social media sites 'like Facebook'. Top of the trust league table was the NHS, followed by banks.

The ranking for trusting organisations with data was almost exactly the same as their ranking for trust in general, with government departments and retailers in the middle. Other factors affecting our willingness to share personal information included knowing the reason why it was required, whether it would be shared more widely, and how securely it would be stored.

I would describe all these as aspects of a trust relationship. Being more concerned about governments having your data, or big corporations trying to sell you stuff, reflects your general feelings of trust towards them.

Personally, I'm relaxed about data-driven advertising. It seems to struggle with my profile. Today, Facebook's offering me cycling T-shirts (nope) and advice for unplanned pregnancy (nope). Twitter is giving me stylish men's shoes (still nope) data software (better) and wine (now we're talking). But relevant or hilariously wrong, nobody forces me to buy something by showing me an advert.

However, since I started writing this book I've become more wary about downloading apps on to my phone and allowing them access to my contacts list or location. I check what I'm agreeing to before I start using the latest free timesaver.

Technology can help guard privacy as well as invade it. It's a fast-changing subject, but I've put a few tips into an appendix in case you're interested.

But more important, in my view, is to have very public conversations about privacy and why it's important. Unless we regard having a private life as a key part of being an adult in modern society, we can't defend it.

Big data makes it much easier to erode the distinction between public and private, but it's not technology that decided the solution to falling trust was more transparency, and the answer to social atomisation was to ask the authorities to watch over our neighbours.

Back to DAC

Let's go back to Oakland, where I've given up trying to talk my way into an office to see a row of computers, and returned with Brian Hofer to the City Center. At a table outside a cafe, a private conversation in a public space, he tells me the whole story.

In July 2013, not long after the Snowden revelations started to emerge, a local man called Josh Daniels noticed a routine item on the city council agenda, a vote to approve Phase 1 of the DAC, a surveillance system for the port. He got a few people along to the council meeting.

'The first speaker was Josh Smith,' Brian recounts, 'and he says: "Where is your privacy policy?" Everybody just went quiet. The vote was still unanimous to proceed, but it got the community aware of the thing. So both those Joshes and some other members were the founding members of Oakland Privacy Working Group.'

The new group immediately exercised their rights, submitting a Public Record Act request for all the DAC documents. Making them public would help reach a wider audience. Including Brian.

'I had no idea. I pay attention to politics, I read the newspaper. I had no idea this was happening until I read a December 2013 article in the *East Bay Express* which analysed a lot of the public record documents. By that point the project was already six months old. The very next day was an Oakland privacy meeting so I just showed up and, you know: How can I help?'

By this time the cameras and cables were installed in the port, and the council was preparing to approve Phase 2, installing equipment in the city and software to link everything together.

'So we showed up to the Public Safety Committee and that's when Oakland Privacy first threatened to sue the City of Oakland. That generated a great deal of press obviously and got the council's attention. It opened the door to letting us start educating them. They had no idea what the system was and what it could do.'

That committee passed the hot potato on to the full council meeting, giving Oakland Privacy time to build a coalition with the American Civil Liberties Union (ACLU) and more than a dozen other organisations. More than 100 members of the public spoke against the DAC at the February council meeting, and the council postponed the vote to March.

'That was the first real clue that we were on to something, that we had a chance to turn around two previous unanimous votes to proceed.'

It also gave them another three weeks to build support among 'left, right, centre ... well, you know, we don't really have a right in Oakland, but those not so progressive, that were concerned about taxpayer costs.' So by the time of the meeting, 'We had 45 organisations, 200 public speakers showed up, the city council meeting started at 5.30 in the evening and the vote didn't happen until 1 in the morning.

People spoke, unanimously again, opposed to the project, and the city council voted to return it to its original 2008/9 status when it was just port infrastructure. They got rid of facial recognition, automatic licence-plate readers, they removed the city portion from the project, prohibited retention of any data.'

Within months, the campaign had convinced the city council to abandon the planned panopticon. But, says Brian, there was even better news, that the campaign didn't anticipate:

'They created the ad hoc committee, the citizens' committee, that would draft the privacy policy to regulate it. Previously the city administration had given lip service: Oh yeah, we'll work on a policy. We just felt they were going through the motions, that they hadn't really decided to address privacy concerns. So giving us that responsibility was amazing.'

Not everybody saw it that way, says Brian, 'On March 4, a lot of activists walked out of there thinking that we lost. They were horribly disappointed, they thought we'd wasted all this time. They thought our council was unresponsive. That is just not true. Not one of those 200 public speakers ever said anything about the tsunami warning system or the earthquake. There's obviously parts of this project that were benign. There's zero civil liberties risk from a tsunami warning reporting system. So March 4 was a true victory.'

Brian sees a changed attitude in City Hall. 'You're now seeing in the resolutions that are being written, in the questions that staff are asking, that they're considering privacy immediately, upfront, instead of way after the fact after an activist yells at them.'

That was in 2014. A privacy policy governing the DAC, drafted with the citizens' ad hoc Privacy Committee, including Brian Hofer and the ACLU, was passed by the city council in June 2015. It includes these words on privacy:

Privacy includes our right to keep a domain around us, which includes all those things that are part of us, such as our body, home, property, thoughts, feelings, associations, secrets, and identity. The right to privacy gives us the ability to choose which parts in this domain can be accessed by others, and to control the extent, manner and timing of the use of those parts we choose to disclose.

Put like this, you can see why your personal data is more important than just a slew of numbers across a spreadsheet.

The importance of privacy can be illustrated by dividing privacy into three equally significant parts: 1) Secrecy – our ability to keep our opinions known only to those we intend to receive them, without secrecy, people may not discuss affairs with whom they choose, excluding those with whom they do not wish to converse. 2) Anonymity – secrecy about who is sending and receiving an opinion or message, and 3) Autonomy – ability to make our own life decisions free from any force that has violated our secrecy or anonymity.

Without privacy, nobody can be fully autonomous or free.

This Policy is designed to promote a 'presumption of privacy', which simply means that individuals do not relinquish their right to privacy when they leave private spaces and that as a general rule people do not expect or desire for law enforcement to monitor, record, and/or aggregate their activities without cause or as a consequence of participating in modern society.

Just because you can collect this information, in other words, doesn't mean that you should, or that we will give you permission to do so. Just because Brian and I were talking outside a cafe, doesn't mean we expect you to come by and record everything we say.

The next stage is a standing Privacy Committee, which will draft a Surveillance Ordinance governing future technologies, based on a model written by the ACLU. And Oakland is now being used as a model for other cities, says Brian.

'I've talked with government groups and privacy activists in other places and they've said: It's all well and good, but it really requires an informed citizenry and active monitoring of law enforcement. And I'm like: Yeah, but that's what we're supposed to do! This is America, we're supposed to hold people accountable, and government is supposed to be transparent.'

'I know in a lot of places that just doesn't happen,' he concedes. 'In Oakland we're blessed that people care and are vocal and passionate and will hold people accountable, so we've been able to generate some political pressure. But it sadly is true that in a lot of places our model might not work, because these policies would get adopted and then just completely ignored.'

So what's Brian's top tip for other cities?

'The first thing is pay attention, watch the agenda, watch what local government is doing.

Number two, you got to get organised. Build a coalition.'

Any city big enough to be getting surveillance equipment is big enough to have active community groups, he says.

'Inform those people, and develop relationships with the elected leaders. Hold them accountable and educate. We were successful at generating a lot of media attention, in part because of the lawsuit, but I think the more effective strategy was educating the people that would be voting. You inform people and they make a good decision. That's encouraging.

It's not just understanding the technology, it's also thinking: Well, if the technology can do that thing we do want, like predicting a tsunami, it could also do this, and follow these people around the city, and make a note of who's been meeting with who, and whose cellphones are in the square.'

Wrong side of town

Using data in law enforcement is not only a problem for individual privacy. It can also have the consequence, intended or not, of targeting groups of people.

A journalist from *Ars Technica* obtained the data from Oakland's automated licence-plate readers and, says Brian, 'put together a map and showed super-high concentration in East and West Oakland. They're not up in the hills targeting the rich white folks, they're in these other communities of colour.'

Brian's not convinced by the police argument: 'Well that's where the crimes happen.' Less than two in 1,000 of those licence plates matched a suspicious vehicle on the police database. 'How come you have a 0.16 per cent hit rate? Where is the crime? If it's not pulling up anything what are you doing?'

Now Oakland has put aside money for PredPol, the predictive policing software we met in Chapter 6. Once again, Brian's sceptical.

'Just up the road in Richmond, their contract was discontinued because it was ineffective. The *East Bay Express* did a write-up on all these different police departments that are getting rid of it because it just doesn't work.'

The *Express* examined crime statistics in several cities where PredPol was in use, publishing their report in June 2015:

'The total number of crimes logged by the Santa Cruz police in 2011, when the police began using PredPol, was significantly above the city's 10-year average. Last year, after three years of using PredPol to predict crime, the city's total number of reported crimes remained significantly above the 10-year average.

Furthermore, in prior years, there were much more significant declines in assaults and other crimes, despite the fact that this was before the Santa Cruz police were using PredPol. The drops in specific types of crime in both cities appear to be just random fluctuations.'

In other words, the apparent success of PredPol's software is down to selectively publishing figures that show improvements. And if you start using it in a high-crime year, crime levels will tend to fall in the next year or so, for the same reasons that earthquakes seem to become rarer after an especially bad year. It's our old friend, regression to the mean.

But that's not Brian's only objection: 'It's going to lead to bias, to racial profiling, because you're telling me to go to a certain intersection at a certain time and there's going to be a crime there. What happens when I get there? I'm going to think all those people are criminals. You've already put this image in my mind. You tell someone: Don't think of an elephant! You just thought of an elephant,' he says.

'So we're going to tell people: Don't suspect Black people in East Oakland! If you're sending me there, and you're telling me this is a crime hotspot based on your PredPol software, I'm going to suspect these people are criminals.'

And, because I am there, if I see any minor lawbreaking, I will make an arrest. The same activity could be going on elsewhere, but if I'm not there to see it, I won't arrest anyone. So the crime hotspot notches up another arrest, and the prophecy becomes self-perpetuating.

'Exactly. You're sending me here for a crime. I need to justify this to my supervisors and to the city council. Now I'm going to start issuing citations for all this this low-level junk no one cares about: He's got an open beer container on the sidewalk. Who cares! You shouldn't have been there enforcing it, and now it's just going to further lead to distrust in the community.

That's the problem, even if the technology somehow magically becomes accurate, community relationships are going to be damaged.'

Privacy is an important issue, and one that won't be resolved by technology alone. But the profiling potential of big data, letting the algorithm predict that my part of town will be a crime hotbed and your children are at risk of dropping out of school, is something I find equally worrying. And that's what we'll look at it in the next chapter.

CHAPTER NINE

Who do we think you are?

Death by ice cream

Shark attacks are more common in months when people buy more ice cream, noted writer Michael Blastland in a 2008 article for BBC online. Then he invited readers to speculate on why this might be so, and got hundreds of responses, including: 'Eating ice cream will cause people to urinate a large quantity of lipids (fats) similarly to seals and other fatty sea mammals, which happen to be the sharks' favourite prey', 'It is easier to swim away from a shark if you are not trying to hold on to an ice cream', and 'Sharks will stop at nothing for a 99 with a Flake'.

Some, missing the joke, pointed out the real reason: both increase in hot weather.

Finding a correlation between death by shark and buying ice cream does not mean that there's a causal link between the two. Banning ice cream vans would not save lives by reducing the number of shark attacks. Conversely, reducing shark-related deaths probably wouldn't reduce ice cream sales. If anything, feeling safer might attract more people to the beach and boost ice cream consumption.

A few years ago, in the first excitement around big data, some people claimed that, in the future, correlation would be enough. Just as the striking link between smoking and developing lung cancer was enough to change people's habits even before scientists understood how smoking caused an increased risk of cancer,* spotting patterns from massive datasets would enable effective policies without understanding underlying causes.

---

* Though, as we saw earlier, Doll and Hill spotted the relationship between smoking and lung cancer because they looked for it, not because it popped out of a correlation engine.

So local councils, which now have responsibility for public health in the UK, might decide to ban ice cream vans from half the beaches, to see if it did lower the incidence of shark attacks. And if they did that, I think they'd find that the number of shark attacks fell on those beaches and rose on the other beaches, which would strongly suggest that their policy was working. The responsible thing to do, then, would be to ban all ice cream vans from all beaches. And shark attacks might fall right across the district.

Why? Because if my favourite beach suddenly doesn't sell ice cream, I may move to a different beach that's more fun. And the more people are at a beach, the more chance there is that one of us will be eaten by a shark.

I should point out that very few people are attacked by sharks, even in Australia where the warm waters are more hospitable for both sharks and bathers.* So no UK council would get statistically useful information from this kind of study. Beaches with or without ice cream vans would both report zero shark deaths. You're vastly more likely to die of a heart attack than a shark attack.

Worldwide, the average number of deaths by shark is just under six per year. So fewer than one in a billion people will be killed by a shark in the average year. Whereas, of people who die in a given year, around a third will die of cardiovascular disease (CVD): some problem with their heart or circulation, including heart attacks.

Don't panic, you don't have a one in three chance of dying of a cardiovascular illness this year. That's your approximate chance that CVD will be the cause of your eventual death.

---

*I know this because shark attack casualties were one statistic we used in our show, *Your Days Are Numbered: The Maths of Death,* and stand-up mathematician Matt Parker checked the figures. When we took the show to Australia we scarcely had to adjust the UK probability of death by shark attack. Though while we were in Australia somebody was killed by a shark, which made our show slightly less accurate, but much more tasteless.

Your chance of dying at all this year is small, and depends on factors such as your age, sex, medical history, involvement in gun crime, and how often you swim with sharks. In the UK, the average chance of dying this year varies between one in 10,417 for a woman aged 5–14, to one in six for a man aged 85 or over. By far the greatest risk factor for death is age.

I hope that reassures you that you're statistically unlikely to die in the near future, and very unlikely indeed to die from a shark attack. Go ahead, have that ice cream.

Public health officials are not idiots. They know sharks are not a major threat, especially in chilly British waters, and that any link with ice cream sales is about the time of year, not sweet-toothed sharks.

And most people who got overexcited a few years ago and said things like 'correlation supersedes causation', are now calming down and saying they didn't mean we'd *only* need correlation, and hastily looking through the bin for causation. But because correlation is one of the things big data does very well, it's still central to the data-driven view of the world.

Computers, even sophisticated ones like IBM's Watson or Google's DeepMind, are much better at spotting patterns than having leaps of insight. They're often very complex patterns, about how many points are connected in a network, and how strongly, or relationships between thousands of dimensions of millions of variables, but they are still patterns, not theories about how the world works.

Guitars and football

Spotting patterns, and making decisions based on what we've previously observed, is something we all do. When you get on a busy train, where do you sit? Not just window or aisle, but who do you sit next to? Chances are, you get on the train, do a quick scan of the people who would share your seat and make a snap decision about which to choose.

You may not even be conscious you're doing it. Most of us have other things on our minds, which is why we have a subconscious algorithm, a mental shortcut, of the sort that psychologists call a heuristic. While you're busy planning what to have for dinner, or looking at Tinder, you may not be aware that you're going through a near-automatic process of weighing up your seating options.

Some psychologists compare this subconscious process to the way data scientists apply Bayes' theorem, though without the mathematics. You start with a best guess, based on what you do know, and then revise that guess as new information comes along. You can't be completely certain about how the future will turn out, but you need to make a decision, so you choose the best bet. 'Given the information I have, which choice is most likely to make me happy?' Or, more often, 'Which choice am I least likely to regret?'

What information are you using to predict the outcomes of your seating decision? Unless you happen to see somebody you know, you have to jump to conclusions about everybody else on the train, based on a few pieces of instant information, mainly appearance.

I live near Millwall FC's ground. For American readers, Millwall is a London football team, soccer to you, which used to have such a bad reputation that one of their chants was, 'no one likes us, we don't care'. A football fan wearing the colours of an opposing team might even expect personal violence from Millwall fans.

My train into central London stops at the Millwall ground, so on match days I find London Bridge Station full of dark-blue-and-white scarves, jackets and hats. In full voice, the supporters fill the station's echoing vaults with a wordless, baritone chant as harmonic as any Gregorian monks. There's a lot of herding around by police, so it can be quite disruptive of a smooth journey home.

Knowing this, if I get on a train carriage and have to choose between a seat next to three Millwall-blue-wearing men, and another seat next to a sensitive-looking young man with a guitar case, the choice is obvious.

However, that isn't all I know. Having lived here for six years, I've had lots of contact with Millwall fans and found them polite and friendly. When they're winning, or hoping to win, they radiate joyful energy. When they've just lost, I feel a pang for their disappointment. So I'm happy to sit next to them.

I don't have anything against sensitive young men with guitars, though there's a small danger they might try to show me their poetry. But before you infer that he's musical, creative and in touch with his emotions, get chatting, give him your phone number and go on a date, let me give you a new piece of information.

If you were once asked for your telephone number by an attractive Frenchman, who never rang you as he promised, don't be disappointed. If you suspected that he asked literally hundreds of women for their phone number that day, you could be right. You may have unwittingly been part of a psychological experiment by Nicolas Guégen and his colleagues at the University of South Brittany.

They sent out a young man 'previously evaluated as having a high level of physical attractiveness' to ask 300 young women for their telephone number on the street. In every case, he was to say he found them pretty, and suggest they could go for a drink later. He never did call them back, and they never did go for that drink. I don't know whether the researchers ever phoned all those young women to apologise for falsely raising their hopes.

The script was the same for all 300 of the encounters, but the researchers varied one thing to see if it had any effect on his rate of success. And it did. Empty-handed, the attractive Frenchman got a phone number from 14 per cent of the women. When he carried a guitar case, that success rate went up to 31 per cent. His chance of getting a phone number more than doubled, just by carrying a guitar case.

This study was published in 2013, and received international press coverage, including a piece in *Popular Science* by the admirably named Colin Lecher. So my young man on the train with a guitar case may play the guitar, or he may have read an article saying that carrying a guitar case makes him more likely to get a date.

I don't know what he'll do if he gets the date but can't play the guitar. It would sound a bit shallow to confess he was only carrying a guitar case to chat up women. Then again, it would sound a bit shallow to say you only gave him your number because he plays the guitar. And you might not even be aware that the guitar case was a factor, it might all be part of the subconscious process of predicting who you'd like to date, or sit next to on a train.[*]

This is a silly example, but we all do it all the time. If we didn't, we'd never have the time or the energy for the important stuff, such as reading or looking out of the window. We're using correlation, remembering that in the past the guy whose shoes are held on with string turned out to be smelly, or that you had a fascinating conversation with the woman reading the book on mathematics.[†]

Nominal data

What if we can't see somebody? Names provide another source of clues. Because fashions for names change, your first name says something about your age. Nate Silver's data website, fivethirty-eight, offers a guide to guessing a person's age from their name, noting that over half of all Lisas are in their forties, but an Anna, with an enduringly popular name, could be any age.

Remember the BBC-commissioned study of names and voting intentions? Your name may not influence your voting

---

[*] If you're collecting dating tips, they also tried sending out the same Frenchman with a sports bag, which knocked his success rate down from 14 per cent to 9 per cent. So sporty guys, try carrying your gym kit home in a guitar case.

[†] I know, that goes without saying.

intention, or your career choice – let's hope not, for Colin Lecher's sake – but the name your parents chose for you gives some clues about your parents[*] and thus about your own background.

Suppose I write to you, asking if you can give work experience to one of my students. I have three potential interns for you: Lakisha Washington, Eleanor Cadbury and Alex Clark. They all have the same qualifications, aptitude and experience. Which one would you choose? Can you picture the three students in your mind?

This is a thought experiment, by the way. They're all imaginary, I don't have any students. Clark is a name I picked in tribute to economist Gregory Clark, who has researched links between names and life histories. He studied University of Oxford students, compared their names with the UK population, and found that Eleanors were three times as numerous in the Oxford student population as in the population as a whole. Another study found that the surname Cadbury is associated with high social status. These effects may not apply if you're outside Britain.

Before you all change your child's name to Eleanor Cadbury, let's remember that banning ice cream doesn't prevent shark attacks. My own father, whose parents were working class, and who went to a state school in Liverpool, got a scholarship to Oxford. He's definitely not called Eleanor.

All UK universities are very keen to find students from a wider range of social backgrounds, so a state school kid with good grades will beat a stereotypical Eleanor Cadbury with mediocre grades every time. Students with good grades from poor schools have already shown they have something exceptional about them.

But parents who call their daughters Eleanor are also more likely to give their children the benefit of good schools, lots

---

[*] My name tells you that they were expecting a boy and, stumped for ideas, looked through the character lists of Shakespeare's plays till they found one that sounded nice.

of encouragement and high expectations. The name is a predictor of academic success, not because it causes success, but because it's associated with factors that make success more likely.

However, many researchers have found that your parents' choice of name may have an impact in the wider world. In a study done in the other Cambridge, in Massachusetts, researchers Marianna Bertrand and Senghil Mullainathan sent out résumés (CVs) in response to real job adverts in 2001 and 2002. They varied the CVs by quality, by address and by name, choosing names that would clearly signal the race of the applicant.

You may not be surprised to learn that 'White-sounding' names got 50 per cent more offers of interviews. Our hypothetical White applicant would need to send out 10 résumés to get an interview, but our imaginary African-American applicant would need to send out 15. Similar studies have found similar effects in Britain.

So much for Eleanor and Lakisha. What about Alex, how did you imagine her? Because yes, my imaginary third student is also female. But she's read some research showing that changing the name John to Jennifer at the top of a job application makes recruiters less likely to offer a job, or training. That name change also knocked $4,000 per year off the suggested starting salary.

Social psychologist Corinne Moss-Racusin found that professors of biology, chemistry and physics saw Jennifer as less competent than John for a laboratory manager job, even though all other details of the résumé were the same. In case you tend to picture all science professors as male, by the way, this included female professors. They also favoured John.

So Alex has stopped calling herself Alexandra and removed anything from her CV that might give away her gender. Which might confuse a few employers when they meet her and have to revise their mental picture of this ambitious young man, but at least she will have got that far.

I'm not telling you all this to make you feel bad if you instinctively went for Eleanor as your marketing intern, or Alex to help out in your engineering firm. They're imaginary, nobody's feelings got hurt. I'm just trying to show that our subconscious shortcuts draw on a lot of social shortcuts that can be limiting.

They can also be overcome.

Things sometimes feel as though they're changing very slowly, but a century ago I wouldn't have been allowed to get a degree from Oxford University,* or to vote. Now you've read this chapter, if you ever have a real job or internship on offer, you'll probably think twice before you make lots of assumptions about the candidates. You know it's a more important decision than where to sit on the train.

But if Oxford used big data methods to choose their candidates for interview, and the machine learning algorithm used previous success as a basis for selection, then Eleanor Cadbury would do very nicely. There would be strong correlations between name, what school you went to, and graduating from a good university.

The algorithm might go a bit further and use other publicly available data such as social networks. Knowing existing or previous students, or even being related to them, is a good predictor of getting into an elite university. Photos of the candidate playing lacrosse or polo must be linked quite closely to Oxford and Cambridge admissions, as they're uncommon sports outside private schools. And there's previous educational attainment, of course, which is perhaps the only relevant correlation.

You might think this would be a ludicrous system, but employers are already using algorithms when they decide who to hire, that are not so different from what I've just described.

---

* I still don't have a degree from Oxford, but that's not the point. The point is that if I had applied, and been clever enough, and worked hard, I could have got one. Theoretically I still could.

Algorithm says no

A few blocks from Steven Skiena's desk at Yahoo! Labs in New York, I drop in on the Data and Society Research Institute, a light and open 11th-floor space, with meeting rooms called cheeky things like Bias and Panopticon. The Institute isn't about creating new technology, more about thinking through the technology's implications for our shared future. Things such as how the use of algorithms can disproportionately impact the same groups that have experienced discrimination in the past.

Over lunch, I chat to one of the Fellows, Gideon Litchfield, about a problem he calls algorithmic accountability. 'Technology can perpetuate an existing power dynamic even if it was intended not to,' he says. 'One of the ways we see that is increased profiling, whether it's by law enforcement, by banks, whether it's shops that calculate what price to offer you online based on where you live …'

Or employment.

Data-driven hiring decisions don't only rely on scoring what the candidates have chosen to submit, such as a résumé or the results of psychometric tests. They may also use other publicly available data, such as social media posts.[*]

Some of the things that earn you a red flag, according to a 2014 Data & Society Research Institute report, are the expression of opinions that are derogatory towards a 'protected group of people', photos with drug references, sexually explicit material, 'animal activism' or 'At Risk Populations'. No, I don't know how you get allocated to an 'At Risk Population' either.

Other factors that can affect your chances of getting a job are how far you live from the workplace, your likelihood of needing time off sick, and how strongly you agree with

---

[*] Most human recruiters also go online to get a bigger picture of potential candidates. If you need a minute to go and delete a lot of stuff off your social media profile, go ahead, I'll wait.

statements such as, 'When I'm working for a company I take pride in making it as profitable as possible.'

So if you live in a poor neighbourhood with little local employment, or have poor health, or tend to answer questions too honestly, your chances of getting work are now even smaller.

'You end up with very subtle discrimination happening at many levels,' says Gideon, 'which will naturally enhance the existing segregations in society. Because people are slightly less apt at the moment to get credit, or to get jobs, that will all feed into the data about them, which will feed into their abilities to do other things. There's a risk of it all becoming a self-perpetuating cycle if you like.'

Our human mental shortcuts can have the same failings, as I tried to trick you into proving earlier in this chapter. But with algorithms there's an extra problem, as Gideon reminds me:

'If an algorithm is making decisions, and it's a machine learning algorithm that has been trained on large datasets and is a black box, nobody can look inside it and understand the logic of any decision that it makes.

And this is the problem he calls Algorithmic Accountability: 'If this algorithm makes a decision and people feel like they've been harmed by that decision, or it's prejudiced against them unfairly, what or who do they appeal to? Is it the people who designed the framework in which the algorithm was trained? Is it the people who provided the data on which the algorithm was based? Is it the people who are using the algorithm to back up their decisions, the policymakers? How do you show whether an algorithm is biased or not? How do you show it was the algorithm that was responsible for that?'

Some claim that algorithms, if carefully designed, can overcome the unconscious bias of human recruiters, and lead to more fairness, not less. After all, if you don't feed information about race or gender into the model, how can it be prejudiced?

Gideon pulls a skeptical face. 'People think of algorithms as neutral, it's just a calculation, but of course they have biases within them that are the fruit of which decisions they're used on, or what data goes into them, or the way that they're written, the way they're programmed. But those biases are very hard to detect.

People naturally don't think of algorithms as biased, because they don't have biases in the easy relatable way that human beings have biases, so policymakers tend to think of algorithms as infallible. So all those things come together into the question of algorithmic accountability.'

Innocent until probably pre-guilty

Not knowing why you were turned down for a job is bad enough. But some young people in Chicago had a surprise visit from the police without committing a serious offence, and can't ask the algorithm why they were singled out for such pre-crime attention.

'If you hang around people who are getting shot, even if you're not actively doing anything, then you become exposed,' Andrew Papachristos of Yale University told the *Chicago Tribune*. 'It puts you at risk because of the behaviors of your friends and your associates.'

This is the thinking behind Chicago Police Department's Two Degrees of Association system, for which they got a federal grant to work with Professor Miles Wernick of Illinois Institute of Technology. Going beyond 500m (just over 500-yard) squares that are predicted to be future crime scenes, they wanted to know who would be committing those crimes. So they fed a list of names into a computer that used various risk factors to rank their likelihood of future involvement in homicide or other serious crime.

These risk factors included past criminal and arrest records, the police records of friends and acquaintances, and whether any of those associates have been shot. So previous contact with the police is a good predictor of the police paying closer attention to you in the future. I'm guessing that law students at the University of Chicago and unemployed kids on Chicago's West Side are both likely to smoke weed, but that the law students are less likely to get a rap sheet for it.

'The novelty of our approach is that we are attempting to evaluate the risk of violence in an unbiased, quantitative way,' Professor Wernick told *The Verge* in a 2014 interview. 'This is accomplished in a similar manner to how the medical field has identified statistically that smoking is a risk factor for lung cancer. Of course, everybody who smokes doesn't get lung cancer, but it demonstrably increases the risk dramatically. The same is true of violent crime.'

One difference is that committing a crime, unlike getting lung cancer, is an act of will, a choice. It may be a choice in circumstances not chosen by the criminal, or a last resort, but it was still an action, not a disease caused by a rogue genetic mutation.

Another difference is that I can choose whether to smoke. I can't choose whether to come from the wrong part of the city, or to have cousins who have committed murder.

Miles Wernick is not a criminologist. He is a Professor of electrical and computer engineering, and his previous research is in medical imaging, training machines to spot worrying patterns in brain scans.

I don't know exactly how the algorithm chose its list of 420 people who are not guilty, but not entirely innocent, and neither do they. The police have so far refused Freedom of Information requests.

If it uses machine learning, the basic AI that underpins a lot of data analysis, then nobody knows, exactly. It is a black box. Having given the computer a task, to find the patterns in past data and make predictions about the future, all we get is the results.

Defendants in Pennsylvania may soon be asking what's inside another black box, as their state has decided to use a statistical risk assessment when sentencing.

This kind of tool is already used to make parole decisions by assigning a level of risk of recidivism. If the computer says you're at a high risk of reoffending, you don't get parole. Now the machine learning algorithm developed by Richard Berk at the University of Pennsylvania will have a say in whether people go to prison, and for how long.

A similar system, used in Virginia when sentencing sex offenders, was challenged by the ACLU when a 19-year-old man was sentenced to 18 months in prison for consensual sex with a 14-year-old girl. The ACLU challenge pointed out that, had the perpetrator been 36, he would have been given a lower risk score and so escaped imprisonment.

Using factors unrelated to the crime, over which the defendant may have no control, such as age, employment and education, undermines fundamental fairness in the criminal justice system, says the ACLU:

'These are sentences divined from nothing more than statistical correlations. Using the same logic, if the Crime Commission discovered that people who like hotdogs and drive Buicks are more likely to recidivate, judges will soon be giving them longer sentences, too.'

America incarcerates too many of its population, that is something on which many politicians from all sides agree. Some of them claim that using data to decide who is unlikely to reoffend is one way to keep people out of jail.

Nor are human judges free from prejudice. But they are, at least, accountable for their decisions, and can be asked to justify and explain them. An algorithm is accountable to nobody.

Astrology to four decimal places

Predicting the future is another thing we all do, all the time. Will that guitarist ever phone me? Will Liverpool win tomorrow, or should I call my dad now, while he's still in a good mood? Will it rain on Saturday?

Weather forecasting is pretty good in the short term, thanks partly to good mathematical models. Very powerful computers work with twenty-first-century versions of Galton's maps of pressure and wind direction, using many variables at many points in space and time, based on real observations and the known laws of physics.

But after a couple of days, the accuracy of weather forecasts falls away badly. Because weather systems tend to be chaotic, in the mathematical sense of the word, they're hard to forecast, even with a complex computer program. Weather is a deterministic physical process, but the tiniest change in the start settings escalates quickly into massive changes in the results.

Predicting processes that involve human beings adds a whole new level of unpredictability. The difficulty with using big data to predict the future is that mathematical models assume the future will look like the past, or at least follow the same rules.

Malthus assumed that human population would grow exponentially, by repeated doubling, and food in a linear fashion, adding the same quantity every year, until starvation or disease intervened. Or until the poor could be persuaded to have fewer children. Mathematically, population growth must overtake food production and inevitably lead to hungry mouths outnumbering plates of food.

But Malthus didn't factor in any of the improvements in food production that have allowed it to outstrip population growth while using less of the Earth's surface. In the last 50 years alone, the land area needed to grow a fixed amount of food has fallen by two-thirds.

Predictions using computer models often include caveats about assuming that present trends continue. These warnings don't always make it into the news coverage, so careful studies spelling out their assumptions become alarming headlines about the world in 2050: half of us will be obese, we'll need 60 per cent more food, and sea levels will be 3m (10ft) higher. Probably because of all the obese bathers, displacing tons of seawater every time they dive in.

Perhaps we should ban those ice cream vans after all.

It's helpful to remember a distinction between projection and prediction. You can make a mathematical model, and test how well it fits the past by letting it calculate past values from even earlier data. If it gives you values for 1980, 1990 and 2000 that are close to the observed values for those years, then it fits well. You have a model that's shown it could calculate the present from the past.

Now, you can continue this model forwards to calculate projected figures for the future. And if things don't change too much, and if your assumptions were sound, you stand a good chance of your projections being close to how things turn out in reality. So you can use your projection as a basis for prediction. If real life continues to resemble your model, here's how the future will look.

The difference is that a projection is a continuation of your mathematical model. A prediction is you, making a judgement that your model is a good guide to the future.

Suppose you want to know how tall your child will be. You could measure them every month and use that to predict how tall they will be at the age of 25. If you just noted that they grow about 5cm (2in) a month in the first year, you'd project a height of more than 15m (50ft) at age 25.

If you observed that this rate slows down after the first year, you might assume that an inch a month is a better predictor, or even that you need to include changes in the speed of growth, so the height eventually stops changing. If you use your experience of real life, you might include things like the growth spurt at puberty, the fact that adults don't keep growing indefinitely, or that family members tend to end up being similar in height.

You could use an online height calculator. I tried one that 'uses genetics' by extrapolating from my parents' height to predict that I will be 1.63m (5ft 4in) tall. Which is about 10cm (4in) too short. So either it's not easy to accurately predict from so little data, or my dad isn't really my dad, and my mum managed to have an affair with an even taller man with similar sticky-out ears and terrible puns.

To be fair, the website does say it's only an estimate.

It's also a reminder that when Galton first used regression to model the relationship between parents' height and the heights of their adult children, he was making predictions about the population generally, or the probable range of an individual's height.

The best predictor of your child's adult stature is their current stature. Look up their height and age on the UK-WHO charts to get a predicted height. If your four-year-old son is 112cm (3ft 8in) tall, he'll grow to be 188cm (6ft 2in). That's not a guaranteed adult height, but four out of five boys the same height as your son aged four will be within a couple of inches (6cm) shorter or taller than this height as adults. So that range is an 80 per cent confidence interval.

Or you could use my hairdresser's rule, that the height of a two-year-old is half their adult height. And he's just started wearing his son's hand-me-down trainers, so he should know.

I'm not sure why you would want to predict your infant's adult height, apart from the fact that parents love thinking about their child's future. But to plan for the future, we need some idea of what will be happening in a year's time, or 10 or 50. Not just individually, but as a society. How many people will need trains or schools or hospitals? All these things take time to plan, build and staff. And making predictions about them means making guesses about the future behaviour of millions of people.

Luckily, as Quetelet noted, it's easier to predict what a population will do than an individual. The variations cancel each other out a bit, and the underlying pattern is more coherent. Your local bar can't know if you, personally, will turn up tonight, or how much you'll drink, but they can have some idea of roughly how many people will turn up, and overall beer consumption on an average Saturday night at this time of year.

But if there's one thing human history should teach us, it's that things will change in ways that history doesn't predict.

Remember Turing and his little robot child? He imagined a thinking robot, an artificial intelligence with feelings that might be hurt by the school bullies. But he still pictured little robot Turing Junior having to bring in the coal to heat the house, and possibly to cook dinner and heat the water for the laundry. The infamous prediction that the market for computers was 'six' simply didn't foresee that they would become so fast and cheap to make, so compact and so widely useful.

And we may want things to change. A century ago, when women in Britain were sabotaging the census in their campaign to vote equally with men, there were some who said we weren't naturally suited to politics. And the data would have backed them up. Women in the early twentieth century were less educated, and less involved in public life, than men.

Nothing about this would inevitably lead to the conclusion that women, therefore, should not get the vote. You could take the opposite view and say that women need not only the vote, but also equal access to education and greater involvement in public life. Many people did, and those things too have changed since 1916.

Big data gives us a tremendous opportunity to use observations of the past and present, processed with powerful software, to make very detailed projections and help us predict the future. It can also help us prepare for the future in ways we're only just beginning to explore.

But it has some pitfalls that we need to beware of, like a tendency to dazzle us with precision. Even when the data analysts include confidence intervals and other reminders that forecasting the future is a grey art, we look for certainty. It can be comforting to have a picture of what to expect.

It may be a grim picture, of a world in which obese people bob about in ever-rising sea water while the entire surface of the world is one intensive factory farm, but at least it's predictable. Which we often prefer to a future in which everything is uncertain.

We have to make decisions, based on too little information, with unforeseeable consequences. No wonder people are tempted to let a machine make the predictions, take the decisions, and bear the blame.

Average man

I once got a feedback card in the post after training a bunch of engineers. On it was written one word: AVERAGE.

Now I've let you wince in empathy, I should tell you it was a joke, and I laughed, because I am a person, not a machine, and I knew the context. But nobody wants to be described as average, do they?

And rightly so. It's useful to know the average when you're studying a population. Average life expectancy keeps going up. That's great news. But it's no guarantee that I, personally, will live a certain number of years. If I look up the official life expectancy figures for my area, I can expect to live to 84. If I moved across London to Camden I could postpone my predicted death to the age of 87.

How? By joining a population with a better average. I don't know why, but I'm guessing Camden includes more well-off areas, because wealth is a good predictor of being healthier and living longer. I don't think it's the fact that I live in a low area near the Thames, and Camden is uphill, further from the miasma.

It could be that those north-Londoners are thinner, less likely to smoke, more likely to jog, and so on. It's hard to separate those things out from having more cash and a less precarious existence, because wealthier people tend to be thinner, take more exercise and smoke less. On average.

When health researchers look for factors that increase the risk of diseases, they have to handle the data carefully to try and reduce the confounding effect of things that go together but aren't necessarily implicated. Statistical innocent bystanders, if you like, found at the scene of the crime and now on somebody's list.

I'm unlikely to change the date and time of my own death by moving to Camden, unless I step out from behind the removal van with a box I can't see over and am struck by a bus. Knowing the average for my population is of limited use for me, personally. None of us is the average person imagined by Quetelet.

I found a website that takes a few more details to give me a more precise predicted departure date. By entering my BMI, my outlook (optimistic) and how much I drink, I've been given an estimated lifespan of 84 years, one month and eight days. So you'd better all come to my 84th birthday party.

Some of these websites are designed to scare you into putting money aside for retirement. Look how long you'll live! Quick, start saving!* Others are designed to scare you into living a healthier life. Look how soon you'll die! Quick, stop smoking!

Some let you change some of your habits onscreen and see how much extra time you'd get, on average. I discover that I'd probably only lose two years if I started smoking now. And I could probably add two years by taking more exercise, so I could start jogging to the ciggie shop and be quits. Though, as I'd be spending all the money I really should be putting into a pension fund on cigarettes, I'd be looking forward to an impoverished old age.

But again, this is deceptively specific. Smoking, like most lifestyle choices, does not guarantee a particular outcome. All you do is shift your odds. You could have no vices, be unlucky and die young, or be that lucky 104-year-old, lighting a fag from all the candles on your highly fattening birthday cake.

It's useful for the National Health Service in my area to know roughly how many of us will be wanting hip replacements, hearing aids and hover chairs, or whatever we'll be getting around in when I'm old. But it's of limited use to me.

---

* No, I don't have a pension fund. Please buy as many copies of this book as you can afford.

If 84 is my expected age of death, I have a 50 per cent chance of living that long. Doesn't sound so good, does it? But I also have a 50 per cent chance of living longer, which sounds better. My chance of dying in any given year goes up every year from now until ... Well, until it stops going up because it's reached one in one.

To be pessimistic, I could die within the next year. To be wildly optimistic, I could live to 114. Between those dates, it's anybody's guess. So let's stop worrying about how long our lives will be, and start thinking about how we live them.

CHAPTER TEN

Are you a data point or a human being?

I'm sitting outside a cafe in south-east London on a bright, crisp autumn morning with Will Davies. He teaches Political Economy across the road at Goldsmiths, University of London. His smartphone has just sent him 'some new coffee thing', a bit late as we've already bought ours. The digital assistant that anticipates your every need still has some way to go.

He's using Google Now for research purposes, he says.

'I think it's terrible really, but it's interesting. It tries to work out what my interests are, what I might want to read. If I had to be on a plane in three hours it would be telling me: Go now! It would have deduced that from scanning my email. I no longer need to think about things that the machine can think about for me, and then I can presumably focus on something more important.'

Davies puts his phone down on the table. 'But ultimately what interests me as a sociologist is this renunciation of agency, or autonomy, or thought to the machine.'

He also wears one of the many wearable self-tracking devices that collects biometric data, heart rate, motion and so on. It's designed to help him be more aware of his own habits and tweak them to healthier ones. 'It tracks my sleep, it tracks various things ... They say the Apple Watch will detect emotions within two to three years because it will have sentiment analysis algorithms and it will combine those with biometrics.

It's got so much Silicon Valley ideology built into it. It assumes that all I want to do is walk, sleep, drink water and work really hard, so it doesn't really have any understanding of me whatsoever.' He laughs. 'I would like to sleep more, but …' Drinking coffee in the sunshine, he doesn't have the air of somebody whose life revolves around hydration and exercise.

'Its obsession is with things that are quantifiable, with these metronomic aspects of life. They try and make you care about these things, but if you fundamentally aren't that interested there's nothing they can do.'

He looks at the biometric device. 'This thing says to me: Make sure you go to bed early tonight because then you'll be really active tomorrow!

And then it says to me: Be really active today because then you'll sleep well tonight!

It's just trying to get you into this loop of inputs and outputs which is kind of a cybernetic fantasy.'

I don't use wearables, mainly because of a visceral sense that I don't want anyone else, even a database, knowing when my heart beats a little faster. But Will Davies doesn't see his wearable like Big Brother as portrayed in *1984*.

'We hear the word surveillance and it still has Orwellian tones, but it's not the same any more. It's not someone listening to you through the wall, it's not the Stasi or MI5 trying to catch a criminal or a spy. A lot of surveillance now is for our own good.'

And Davies doesn't just mean protection from crime and terrorism. 'A huge subsection of the Silicon Valley industry at the moment is all around mental health monitoring and prediction of mental health episodes. There are loads of apps and wearables aimed at helping you manage your health, predict when a loved one is going to have a mental health episode, predict when ex-army veterans might be developing suicidal thoughts, from their behaviour. This is not about catching bad guys, this is about catching undesirable events.'

Which means monitoring not just your body, but your mind, says Davies: 'The other area that everyone's very excited about is affective computing, reading your mood through your face. There's a company called Affectiva which spun out of MIT media lab …'

Is that the one that claims it can read emotions better than a person?

'Yes, but it can't. One of the issues here is what does "better than a person" mean? In a way it starts from a misunderstanding of what an emotion is. The idea that an emotion is something you can be correct or incorrect about is already a misunderstanding of an emotion.

I could do that –' Will gestures expressively with his free hand, 'and we know what I'm doing. But the idea that there's a true or false meaning of that, is already a computer scientist's view of the world, that everything is codifiable.

The other way in which it differs from the Orwell view of things is that none of it's interested in individuals, it's only interested in patterns. It's about pattern recognition. That's how the machine learning comes into it. What you want is to collect so much data that the computers work out for themselves what counts as a symptom, and cut out professional judgement.'

So the artificial intelligence removes from the expert the responsibility of judging what's best, and the smartphones and wearables remove from the user the need to decide what to do.

Will Davies links the current success of big data to what he calls a new form of behaviourism. Early in the twentieth century, scientists studied animals as if they were little machines, to see if you could alter their behaviour by changing the stimulus, or the environment in which you placed them. Quite often, you could, at least with pigeons and rats.

Later, this approach was extended to human beings. Instead of asking them questions, you just observed what they did under different conditions. Then you could change the conditions, hoping to change the resulting behaviour.

Now we have devices that can both gather data and deliver information back to the user, the stimulus–behaviour–data collection process can become a continuous loop of adjusting the stimulus to get closer to the desired outcome.

'If you've got constant feedback, loops of information going between you, your body, your behaviour and your environment, your environment can be constantly adjusted to enable you to take decisions which are more optimal,' as Will Davies describes it. 'The question of what is a good decision is no longer something that's purely on the level of the psychological, but is achieved in some kind of systemic loop between the individual and their environment.'

Are you quantified?

Individuals don't always wait for an app to observe them and tell them what to do next. Some see data-gathering as a way of taking more control over their own lives, knowing themselves better, and using that to change what they do. And it's not all solitary navel-gazing, there's a Quantified Self community that meets up to share methods, insights and experience. One recent conference had show-and-tell sessions including:

HOW FOOD-TRACKING SUPPORTED BECOMING A VEGETARIAN – As part of my transition to vegetarianism, I found that taking photos of my food was useful. I'll share my findings from over one year of food journaling with photos, how it facilitated the process, and the obstacles I found in manual tracking.

MENSTRUAL CYCLES, 50 CENT, AND RIGHT SWIPES – Curious to understand how hormones influence my life, I spent six months monitoring how my menstrual cycle affected my choice in music, use of language on Facebook, time spent shopping online, and even my Tinder usage.

DRAW A FACE A DAY – For six months, I drew a face to self- report my mood for the day. This inspired and engaged me more than I expected. The faces triggered my curiosity and provided many insights that motivated me forward. Unintentionally, I balanced my life differently by drawing happy smiles every day.

All these people are using behaviourist approaches on themselves, giving themselves feedback to reinforce things they've decided to change, or simply studying how their own behaviour varies with biological cycles.[*] But even if the measure is itself subjective, like assessing one's own mood, they all focus on quantifiable aspects of human life.

Gathering data on themselves, and looking for patterns, are means by which people try to take more control of their own lives. But the very act of self-measurement, of quantifying the self, changes the meaning of what a self is.

In a purely deterministic universe, as Laplace imagined, a self would be merely an observer of events, of the inevitable effects of causes that were also inevitable. There would be no room for choices, and free will would be an illusion. Indeed, some determinists today argue that the self is an illusion, arising from the biological machinery of our brains. To which philosopher Mary Midgley[†] replies crisply: 'We naturally ask just who it is that is being thus deluded?'

This, to me, is the key distinction between the quantified and the quantifying self.

The quantifiable is always a *what*, something that can be measured, weighed, in some way reduced to digital data that a computer can collect, store and analyse. And yes, each of us can play that role in many datasets. I can be weighed, my movements mapped, my heart rate tracked over a week and plotted against my Tinder usage or goals scored by Liverpool FC. I could be quantified, in the sense of measuring distinct dimensions, aspects of my existence.

But the quantifying self is a *who*, a person who decided to find out more by gathering data for a purpose, who acts in order to achieve some goal, whether that is becoming a vegetarian, feeling subjectively happier, or just satisfying curiosity.

This autonomous self cannot be quantified.

---

[*] I'm not sure how you'd use that knowledge, except perhaps to avoid Tinder on days of the month when you're swiping indiscriminately.
[†] In her book, *Are You an Illusion?*

I love watching TV shows about archaeologists who dig up human skeletons and, by analysing the bones, tell us that this person lived with chronic toothache for 50 years, or survived a sword wound but had a scarred hand and head for the rest of their life, or gradually lost their nose to syphilis. It always reminds me how lucky I am to be living today, in an advanced country with modern health care.

But however much they can discover from scientific analysis of somebody's skeleton, it tells us nothing important about who that person was. They could have been Queen Elizabeth I of England, US President Andrew Jackson or astronomer Tycho Brahe. They could have been the first person to brew beer, or think of Pythagoras' theorem. Or they could simply have lived a normal life of routine work and family relationships, every bit as full of meaning as yours or mine. With each of them, an entire universe of thought and experience was extinguished at the moment of their death.

Looking at my bones when I'm dead won't tell you anything of value about who I am, and neither will downloading my data while I'm alive.

Look smart!

For Will Davies, the view of the world, and of ourselves, founded on data analytics and behaviourism, 'assumes a kind of shifting of autonomy, away from the self towards some centralised form of expertise'. But why is it happening now? Is it just that the technology is so advanced?

'I think partly it's a thing that's been around for a while,' he says, 'a breakdown of ideologies and political authorities who can offer a narrative about society, about our social lives or collective lives.

Often the idea that human beings can be entirely understood purely through observation and data points arises during periods of moral and cultural crisis. Today, everyone thinks we live in a society of free markets, and we do to some extent, but I'm not sure anyone really believes any longer that is a basis on which to organise all of our social life or political life.'

The financial crash, and the worldwide recession that followed, certainly rattled any belief that free markets alone could be relied upon to keep the world running. But with no real political alternative on offer, where else can our leaders turn?

'I think in a way behaviourism plus data is offering a new technocratic solution. Which is appealing at the level of government, because from a policy perspective it's the complexity of problems that leads to the appeal of some of this stuff. The promise of behaviourism is that we can solve complex problems by collecting lots of data.'

Data-driven solutions like smart cities, for example, promise to resolve problems by connecting things, and people, in one continuous, self-adjusting system.

'When democracy breaks down, smartness is an alternative to democracy,' says Davies, 'because smartness says: What would be a legitimate authority over the city of London? It's not really possible to imagine because it's too complex, and there's too many people living here, and nobody believes in politics anyway any more. So what we'll do is capture all the data, and the machines can work it out.'

Then he puts into words my own nagging fear: 'The ideal of smartness is social life cleansed of all the difficult bits, of negotiation, of listening to each other, of discovering that we want different things from each other. All the things that politics and democracy confront, and are often difficult and painful, the ideal of smartness is to take them away.'

Smartness works on the basis that we are not very smart, that data knows how to run things better than we do. That, in some cases, it's humans and our behaviour that are the problem.

'It's this idea that there's nothing about human life that is ultimately any different from pigeons and rats.'

Davies contrasts this behaviourist approach to a different philosophy: 'Aristotle said man was a *zoon politikon*, a political animal. Human beings have two traits that other animals don't have. One is a tendency to distinguish good from bad, in a moral sense, and the other is a tendency to speak to each other.

Take these two things together and you have this thing called the polis, which is where men would go out and debate the nature of the good life, and try to take decisions based on those discussions. That's the lifeblood of democracy, even in its crappy forms, a debate about the nature of the good which is spoken and listened to.'

This vision of human life as a debate about what a good life means, between people who then try to act on what they have democratically agreed, seems to me fundamentally at odds with letting big data run our lives for us. Where is Aristotle's polis in the smart city?

This grimy south London street, on which we have to raise our voices above the scream of lorry brakes, and have our conversation interrupted by lost drunks, is far from most visions of a smart city. But as the yellow leaves blow by and the tang of smoke from the next table pricks my nostrils, it's infinitely closer to my ideal city than a clean, quiet metropolis optimised by a network of connected devices.

Data-ocracy

Can we use the abundance of information that big data offers to rebuild democracy on new foundations?

A few years ago, I chaired a debate on evidence-based policy, in which David Willetts, then UK Science Minister, argued that in today's inchoate political world, evidence could be a common starting point.

'As our society changes and we become more diverse with a greater range of moral and religious views, evidence matters increasingly,' he said. Working in a coalition government of Conservative and Liberal Democrat parties had thrust evidence-based policy to his attention. 'When you're trying to agree a way ahead, and you can't assume tribal loyalty, you find yourself drawing on evidence about what works, what doesn't work. That's the backdrop which I think we can share and accept.'

We may no longer have a shared language of left and right, a common recognition of where the dividing lines fall in society, or an instinctive feel for what follows from loyalty to a few major ideas: equality, liberty, state protection for the weak, or a free market to encourage growing wealth. But we can all sit down with some figures, graphs, a chunk of data, as the building blocks of a shared road to the future.

You might be thinking this shows a naive faith in the objectivity of facts, and of our willingness to accept them uncritically. One unfortunate consequence of the drive towards evidence-based policy is the rise of policy-based evidence, facts and figures marshalled purely to back proposals that politicians have already planned.

Which is understandable. Politicians are among the least-trusted members of society, right down at the bottom with journalists in most polls. Politics itself is viewed with suspicion, as if the very claim to have beliefs about how society should change is somehow morally dubious.

Evidence can be an unsullied champion in the dirty world of politics, a factual Joan of Arc, leading your army to victory with its sword unbloodied and virtue intact. Which would make big data … Joan of Arc in a giant robotic exoskeleton? Still pure, but now with unstoppable technological power. Though, as I hope I've convinced you, we should regard all data with a critical eye, even if we don't suspect it of being a cynical ploy to dazzle you with digital infallibility.

You may also be muttering that there is more to politics than knowing 'what works and what doesn't work'. Seeking a policy that works sounds very sensible, but it presupposes that we have agreed on our end goal. Capital punishment has 100 per cent success rate in preventing reoffending, but we don't all agree on using it just because it works.

I should add that David Willetts advised us not to be naive or uncritical, and to remember that 'even rigorous evidence doesn't always reach the same conclusion', and of the 'yawning issue of the is/ought dilemma'. Knowing what a problem is doesn't tell us what we ought to do about it.

'We're not a technocracy, we're a democracy,' he declared, 'and a bloody good thing we're a democracy.'

Politics should be founded in as true a picture of the real world as we can build, of course. But it should not be limited to that.

No movement for social progress, from the French Revolution to the Civil Rights Movement, was an evidence-based policy. It would have been far easier to look for evidence that existing social inequalities were founded in natural differences, as many did and some still do, than that freedom or equality are worth fighting for. Those ideas look to a vision of a society existing only in the imagined future.

Trust

As I've learned more, through writing this book, about how much of my data is routinely collected, stored, shared, linked together and analysed, I've become more circumspect. Instead of automatically ticking the box, I now read the small print, weigh up the benefits and the possible consequences. If I can achieve the same thing without handing over my personal data, I will.

I think we will gradually move towards a different approach to data, and see it as something over which we should retain some control. I hope so. And I hope we get there, not in reaction to hackers getting unauthorised access, but because we value our privacy and our autonomy.

But I am not what's sometimes described as a 'data fundamentalist'. I see the value in sharing data, both for my own convenience, and for the good of wider society.

It comes back to trust. At the moment, the status quo is that our trust is taken for granted. Our most personal information, our private exchanges, our network of friends, are used by others without our consent. Or we give consent in a formal way, without having much idea of what it means.

In some cases, we trust the organisation in question. Two-thirds of people surveyed by the DMA in 2015 had a high level of trust in the National Health Service, both generally and when it came to their data.

If that trust is abused, that picture could change. Focus groups interviewed for the Wellcome Trust in 2013 expressed specific worries about the prospect of having their health data linked with other types of data. They foresaw a changing NHS, with more limited budgets and a greater emphasis on prevention through behaviour change, refusing certain individuals medical treatment because of information about their lifestyle or non-medical activities.

Given that some NHS treatments are already rationed on the basis of patients smoking or being overweight, this is a reasonable worry. We could be profiled as *at risk* of poor health outcomes, and offered non-negotiable support, as a condition of future medical treatment.

In the US, where health care is provided through insurance often paid for by employers, there are already moves to encourage employees to wear Fitbits and similar self-monitoring devices.

Responsible employers helping their staff improve their own health through exercise and diet, no doubt, and a way to collect data for health research. But medical insurance companies, or employers themselves, could also track who is jogging and who is sitting down for a fried breakfast, outside work hours, in what used to be an employee's private life.

So trust is not only about taking care to avoid breaches of privacy, but also about the uses to which data might be put in future.

We need to trust each other. I need to be able to tell my doctor the truth about what I have been doing, feeling, smoking, in order for him to look after my interests and give me the best treatment. If I have to go into hospital, I want my medical records to be available to the emergency doctor so she doesn't have to waste valuable time asking me questions. Assuming I am in a fit state to speak. One of my off-the-record conversations with somebody working in health-care data, who trusted me not to print his name, ended on the chilling note, 'people have died with their privacy intact'.

Much of human civilisation runs on trust. We trust that the trains will run, that the bank will look after our money, that people in general are more likely to help us than harm us. Sometimes that trust is betrayed, but it is one of the things that ties society together.

But trust isn't a bankable resource, weighed like grain or poured out like milk. To trust is an action, a relationship between people, or organisations. It has to be freely chosen, not assumed.

Trust isn't binary, an all-or-nothing relationship. I'm happy for my bank to know my financial history, but I wouldn't want it reading my diary. Indeed, I have friends whom I'd trust with my life, but not my personal secrets. And though we always risk betrayal when we decide to trust, we don't do so blindly. We weigh up what we know, invoke our past experience and the judgement of others, before we trust.

Too often, when our trust is taken as implicit instead of explicitly requested, it's because we don't feel inclined to trust the company or organisation that wants our data, and they know that. But would we not be more willing to trust if we could decide what data to share, how long it could be retained, and the purposes to which it could be put?

As we become more savvy about what data is gathered, how it may be used, and its value to those who gather it, I hope things will change. We may force this change, as we start to use technology or simple cunning to hold back more data about ourselves, and demand a better deal before we hand it over.

Who is watching whom?

The other thing about trust is that it's a two-way process. Too often, those who ask for our trust do not reciprocate it.

Many people have pointed to the hypocrisy of governments who want us to acquiesce to mass surveillance while trying to reduce their own accountability through Freedom of Information acts and the like. If we are to be watched and eavesdropped upon by the state, should we not by default be able to access data collected about us, or about the activities of the authorities themselves?

I'm not talking about extreme scenarios, here. Unfortunately, I don't have to be a paranoid conspiracy theorist to see that the intended uses of my data go far beyond screening out terrorists, as described in Chapter 8.

In the name of combating extremism, government agencies take an interest not just in our actions or expressed intentions, but in our opinions and feelings. UK Schools are offered software that monitors pupils' internet activity for trigger words such as 'Jihadi bride', so they can refer their own students to the *Prevent* programme, thereby ensuring they go on to the security services' database.

Is there anything more guaranteed to drive genuinely worrying views underground, away from challenging debate, than knowing you could be on a list for voicing doubt? We are now in the territory of having our thoughts policed.

And not trusting us to weigh up controversial views without turning to murder reflects a very low opinion of us all. The roots of the idea that private conversations are inherently suspect lie in the view that, unsupervised, we are either too stupid or too evil to resist stupid and evil ideas.

However, we don't have to look into such dark corners to see how little we are trusted. Public health researchers spend time and money gathering data on our habits, and linking them to our medical data to find associations between eating takeaway food, for example, and being obese, developing diabetes, and dying earlier than average. Where they find correlations between consuming more takeaway food and dying earlier, they want to take action. So local councils are encouraged to reduce the availability of takeaway food in areas where the population is at risk of obesity and other bad outcomes.

This policy proposal rests on a number of assumptions.

First, that there is a causal relationship between number of fried chicken shops, for example, and rates of obesity. That it's not another case of ice cream vans and shark attacks.

Second, that the takeaways cause the obesity, not the other way around. Could it be that fat, unhealthy people are simply too large to get near the stove and cook their own dinners? OK, that does seem unlikely, but poor health can be a cause of obesity as well as an effect.

Third, that the fat people, or people at risk of becoming fat, don't realise that fried foods are high in calories. Or they do realise, but are unable to resist the tasty temptation of having so many takeaways nearby.

And fourth, that being healthy and not obese is an important goal. You may feel that goes without saying. Wanting people to live longer, healthier lives is a noble goal. But by deciding that, for their own good, you're going to restrict their opportunities to buy takeaway food, you are admitting that you do not trust them to decide what to eat, or how to feed their families.

And by 'them' I mean 'us'.

It's precisely what Will Davies described as the shifting of autonomy away from the self and towards centralised expertise. Closing down fried chicken shops doesn't seem like a very hi-tech example, but such policies are justified by research linking health data with other types of data.

Just as Quetelet looked for external factors to help explain different rates of crime, or of marriage, today's policymakers still try to create circumstances in which we can all live better lives.

Meaning, usually, healthier lives. Our surroundings are described as obesogenic, giving rise to obesity, for various reasons: too many fattening foods, too tempting and cheap and easy to obtain. Too few opportunities for exercise. So a few redesigns might change the environment in which we make our little, everyday choices, often without thinking. It's an approach, drawing on behaviourism, often called nudge theory.

Put the elevator at the back of the building and the staircase at the front. Ban chicken shops from poor neighbourhoods, and ice cream vans from beaches. Or encourage us all to wear a Fitbit or an Apple Watch that will count our steps, calories and inches, and motivate us to increase, decrease and lose as appropriate.

Yes, you may be saying, that is exactly why I got this gadget in the first place: I want to be fitter and, knowing my own laziness gets in the way of my long-term priorities, I'm using data-gathering to motivate myself in the short term. By outsourcing small decisions to my data-driven assistant, I have more time for the important stuff.

Believe me, as I sit at my keyboard, listening to the rain, accumulating inches around the waist instead of steps around the block, I too acknowledge my frailty, the gap between desire and consummation that makes me wish for a machine that will be PA, housekeeper and personal trainer rolled into one. If I gave such a gadget a better-than-evens chance of turning me from an out-of-shape, disorganised writer to a fit, healthy, productive paragon of efficiency, I would be buying one online right now, without having to leave my desk. I have nothing against using technology, psychology and anything else you've got to help each of us attain our goals.

But there's a difference between you or I deciding that we want to run a marathon, or swim the channel, or just be able to run for a train without getting out of breath, and somebody else deciding that we should be nudged so they can hit their population health goals.

Whose good life?

It's fine for a public health professional to say, 'let's set ourselves the target of a healthier population'. It would be weird if they didn't. The problem is, there's a very big difference between the link from action to outcome on a population scale and for me as an individual.

It's useful to study things on a population scale, and to combine different datasets to find hitherto unsuspected links. But it's dangerous to start seeing us as lab rats, behaving in predictable ways as the experimenters change the design of the maze.

We've come across eugenics a couple of times in this book, first when the early statisticians linked the nascent science of genetic inheritance to their ambitions to engineer a better society, and later when genetics, in the big data age, gives us the power to redesign human genomes at an individual level.

I don't worry that we're heading for a world in which genetic outliers are eliminated. So long as we keep choices in the hands of parents, I think the combination of family love and the variety of human tastes and ambitions will protect us. I also think our genes are not the most important thing about us, and certainly don't determine our entire lives.

What worries me more is a new form of determinism. Not race or genetics, but being *at risk* could turn any of us into a problem that must be solved. And the more we appear as data points in a complex model of society, the easier it is to see us only in terms of dimensions that must be optimised to raise the average health, happiness or well-being of the nation.

But, just as trust is an act, not a commodity, risk should be something we decide to take, not a miasma that clings to us like foul air rising from the Thames. Being at risk of something is a passive state, as if becoming a criminal is something that just happens to a certain proportion of youths from low-income backgrounds, or getting obese enough to threaten your health* is something that falls on you from the sky.

---

* Which is more than just a bit of padding. On average, people categorised as 'overweight' have better health outcomes than people with a 'normal' BMI. Most people need to be well into the 'obese' camp before they have a health problem.

I want to go back to one of our assumptions earlier in this chapter, that being healthy and not obese is an important goal.

Across a population, I can see it's a worthwhile goal. And not just for instrumental reasons, such as having a population who could fight a war, or be a productive and profitable workforce, or to minimise health-care costs, though those will all* be in somebody's priority list somewhere. If you're a public servant, and you want to maximise the well-being and welfare of the population, keeping as many of us as possible alive and healthy is a good start. Nothing like being alive to keep your options open.

But it is only a start, or a means, not an end. What is the purpose of life? I doubt your answer would be: To keep breathing as long as possible, with a resting heart rate of 60 beats per minute.

People can have bad reasons to overindulge to the point of damaging their health. They may be self-medicating emotional or existential pain. They may despair about their own future, and therefore see no reason to live for anything but the present. These reasons are unlikely to be susceptible to behavioural nudges.

There are also good reasons to disobey doctors' orders. Many of us drink more than the recommended one or two drinks per day because alcohol strengthens social bonds, which are more valuable to us than sticking to government guidelines. We may indulge in risky activities because they challenge us and make us feel more fully alive.

A lot of the most worthwhile things in life involve taking some risks. Having children is risky. Driving to work is risky. Cycling to work is far more risky. Then again, staying home alone puts you at risk of loneliness which, according to some studies, is as risky as smoking. Minimising risk could become

---

* OK, maybe not the one about fighting a war. Half a dozen people who can sit in a container and fly a drone will probably be enough for that soon.

a full-time job in itself. Is that what you want on your gravestone? 'Lived to 113 by systematically minimising risk'?

Galling though it may be for health professionals, we do not live our lives for the benefit of our doctors. Wherever you feel the meaning and purpose of your life resides, I sincerely hope it is not in longevity per se, or in anything that could be captured in digital form. Your self, in the most important meaning of the word, is not quantifiable.

Too big for its boots

I asked you earlier not to be too easily impressed by the bigness of big data. Or by the power of its computers, or the awe-inspiring mystery of its techniques. Then I proceeded to give you an incomplete tour of what it can do, and should be able to do in the foreseeable future, some of which is very exciting indeed. So I may have polished up the shiny technology to dazzle you still further.

The appeal of big data is not entirely about what it can do for us. As well as the undoubted power and promise of combining new ways to collect digital information with new ways to analyse it, big data has arrived at the right place, at the right time, to capture our imagination.

Because machines collect, sort and analyse the information, the whole process appears to be objective, free from human bias and weakness. As I hope I've shown, this isn't necessarily the case in practice. Human beings decide what to measure, what to look for, and how to interpret or act upon the results. Even where machines are left to look for patterns, writing their own rules as they go, so no human knows the process that led to the outputs, the model is based on past patterns. It therefore has the bias of history woven through it.

The apparent completeness of the datasets, the 'n = all' claim that the sample represents the entire population, adds to the illusion that big data is all-knowing. American writer Christine Rosen, when I interviewed her in 2013, characterised it as a modern oracle.

'When people talk about big data, it's almost spiritual. Everybody will be more healthy, we'll be safer, we'll be risk-free. It's a kind of faith in information that is compelling in some ways, but it's also dangerous. It frees us from asking questions about what we're looking for in the data,' says Rosen.

'Big data is very good at giving us the *how*, but not the *why*. Many of the pressing questions we face are *why* questions. Big data can help us answer some of these questions, but it's not a panacea.'

Christine talks about it filling, 'this need we have for control, explanation. But it's only as good as the judgement of the people analysing the data. We see this amorphous big data force as bringing us wisdom, but judgement is still very important, and that has to be human judgement.'

And yet it continues to exert an authority that is invoked by businesses, governments, scientists, aid organisations, even sports teams. Why is this?

Like all the best oracles, its working remains shrouded in mystery, which means its answers cannot be questioned. And the less faith we have in our own human capacity to understand the world, the greater the wisdom of the machines appears by contrast. If we no longer trust ourselves, our selves, to discuss what is a good life and make reasonable decisions that move us towards it, who are we to trust? Who, or what?

Over cocktails in an American hotel, I got chatting to somebody from a data science company, who told me how he first got interested in using artificial intelligence to spot patterns in data.

As a child, he had travelled some of the politically unstable regions of the world with his parents. His father worked in intelligence, his mother was an artist. In the 1990s, as the world of geopolitics began its seismic shifts after the collapse of the Soviet Union, he wrote his university thesis on the emergence of jihadist movements in the Maghreb.

For him, big data brought together what he'd learned from both his parents. From his father, the urgency of spotting patterns in chaos, finding meaning and predicting change.

From his mother, a love of pointillism. Close up, the painting is a jumble of coloured dots. Step back far enough, and there it is: the fashionable ladies of Paris enjoying a Sunday stroll on La Grande Jatte. The whole picture, so clear you can't imagine not seeing it.

Thirty years ago, understanding the dynamics of global conflict was simpler. Everything hinged on the Cold War, on the ideological opposition between two superpowers. Now, things aren't so clear. Even in domestic politics, left and right are no longer adequate to map out where people might stand. Sometimes it's hard to see any purpose or reason underlying events.

I could see why the dream of machines that put all the data points together until the whole picture leaps out is so appealing, now more than ever. But its ability to answer the *how* questions could be a dangerous distraction from the *why* questions.

I am with Christine Rosen on this. Without human judgement, big data is less useful than a wolf bone with 57 notches. At least a dog can gnaw on a bone.

It takes human judgement to decide what questions to ask, what data to collect, and how to interrogate it. Then it takes more human judgement to interpret the results and decide what actions, if any, should be taken.

Not big enough

In some areas, the claims made for big data are wildly overblown.

The more human beings you study, the more patterns will emerge that you may be able to usefully apply. But to say that a computer, however powerful, can understand human beings better than another person can, reflects a reductive view of humanity, more than a realistic assessment of what machines can do.

If you want to know what people are like, or to predict what any individual will do next, you would do better reading some novels, or watching a Shakespeare play. Or even, if you dare, talking with that person.

In other fields, the potential of the technology is being squandered on unambitious goals. Too many of the projects I have described are about aggregating very small gains. Save a litre of fuel here and there, turn down the central heating by a degree, spare me two minutes of form-filling next time I replace my driving licence.

By adding up lots of small gains, the overall difference can be significant. But instead of expanding the scope of what we can do, they're mostly about doing the same more efficiently. Smart is generally shorthand for doing the same with less.

The preoccupation with changing human behaviour sits well with this. If the goal is to reduce the impact of human beings upon the world, then people are seen as the problem, not the solution. The response to energy shortages is to get people to use less, with gadgets that nudge them to reduce consumption, or by cutting them out of the loop altogether. Let the Internet of Things decide when to put the laundry on, or the television off.

I hope I haven't made you cynical about the promise of big data. I couldn't have written this book without it. From sophisticated internet search engines to transport planning that helped me travel around the world and talk to people doing mind-expanding things, I have harnessed its power, in a small way, to give you a glimpse of its potential.

Remember Ada Lovelace's warning about the tendency, 'first, to overrate what we find to be already interesting or remarkable; and secondly, by a sort of natural reaction, to undervalue the true state of the case'.

We should not let our existing preoccupations, to which big data may not be a good solution, distract us from its real potential.

I would like to see big data flexing its muscles and addressing some bigger problems. Instead of reducing demand, how about increasing supply? Harnessing the analytical power that found the Higgs boson to solve the engineering problems of nuclear fusion could provide us all with plentiful, clean energy. Then we could leave our central heating up as high as we like.

Transport is a similar picture. A smarter transport system that lets us flexibly use a mix of trains, cars* and whatever else, with real-time schedules, flexible payment systems and capacity that responds to demand would be fine. Better still, a system without delays and breakdowns, with trains that are never too full to board. Best of all, a swarm of self-flying cars, with air traffic control through a big data network, that flies each of us directly from A to B with great views on the way.

What holds us back is not the magnitude of these challenges, but a lack of collective vision. In his lovely book, *Profiles of the Future*, published before I was born, Arthur C. Clarke characterises two failures when looking to the future: failure of nerve, and failure of imagination. I'm not sure it's possible to separate the two. Today, when our sense that we can make the future better than the present or past is weak, it is difficult to imagine how we might harness today's technology to reach such a better future.

It's not that big data is too big, it's that it's not big enough. We need more projects like Eamonn Keogh's ambitious schemes to capture the world's insects on a global, public database. Big data could give us unprecedented power to control them, to protect human beings from hunger and disease.

It's not that we need less trust in machines. It's that we need more trust in people.

We need more trust in the people who could use the considerable power of big data as a tool to better understand the world, if they're prepared to use their judgement and be answerable for their decisions. We need to trust those people, not blindly, but as adults also taking responsibility for our decisions.

---

* Self-driving if you like.

We can take charge, far more than we do, of what we choose to share or to keep private, with this organisation or that company. But our individual choices are only part of the challenge.

We also need to be prepared for a clash of ideas and priorities, to argue about what questions we want the powerful machines to help us answer, and how to act on those answers.

For all this, we need more trust in the great mass of people whose wisdom, uneven and fallible as it is, far outstrips the smartest computer when it comes to deciding how to live.

One of whom is you.

I'm not a data point, I am a human being. And so are you.

CHAPTER ELEVEN

Even bigger data

As I predicted, since this book first went to press, big data has only got bigger and stuck its digits into more aspects of our lives. So for this paperback I've attempted a quick roundup of some of the most important new ways in which big data is changing the world, or in which people are changing the world, using big data.

The LHC now pumps out even more data than I described in Chapter 5 – around 1GB per second from CMS alone – and around 50 petabytes per year altogether. Using people's social media accounts to decide whether they should get a loan, as we saw in Chapter 4, has entered the mainstream, with companies like Kreditech and Lenddo selling their services to major lenders. Virtual personal assistants are now everywhere. Amazon's Alexa has taken up residence in many homes, although the humans in the family aren't always in control of the AI's helpful initiatives; not only can a little girl's chatter incite Alexa to order toys online, but a television presenter reporting that very story in the background can also provoke repeats of the over-enthusiastic order.

And though I'm writing this new chapter only a few months before the paperback edition appears, it'll be out of date again by the time you read this page. Nevertheless, here are a few updates touching on what I think are the most important recent developments in big data.

Doctor Watson

Early in 2017 I caught up with IBM Watson, now applying Watson's AI powers to a wide range of sectors, including helping the oil and gas industry to refine its gushing well of data, providing financial advice and creating chatbots to interact with customers. But I was visiting for BBC Radio 4 series *FutureProofing*, to find out more about IBM Watson's role in healthcare.

Building on Watson's capacity to take in information from diverse sources, then search through it for relevant answers and select the most likely solution, IBM has developed several tools that doctors can use for closely targeted patient care.

Cancer doctors can combine faster, cheaper genetic analysis of tumour tissue with Watson's analytic powers. By comparing the patient's results with other samples, and with the latest studies, Watson narrows down the potential treatments to a shortlist. Doctors can read recommended options, including a summary of the good and bad points of each, and references to the research that led Watson to make a particular recommendation.

Other medical services include matching patients with clinical trials for which they are eligible, and monitoring for 'adverse events' or unwanted side effects. In each case, the ultimate choice remains with the human doctor, working with the patient and their personal priorities.

Google's DeepMind AI division is also not content with beating humans at their own games.* Like Watson, DeepMind is turning to medical investigations.

One partnership with London's Moorfields Eye Hospital uses anonymised patient data to look for patterns in eye scans that might help the diagnosis or treatment of macular degeneration or diabetic retinopathy. Another London partnership, with UCL Hospitals Trust, analyses scans of head and neck cancer patients, aiming to improve the targeting of radiotherapy.

London's Royal Free Hospital approached DeepMind to develop an app called STREAMS, designed to alert specialist doctors to fast-developing conditions in hospital patients. In early 2017, doctors started using STREAMS to detect a kidney condition called Acute Kidney Injury (AKI). 'The use of slow and outdated technology means that important changes in a patient's condition often don't get brought to the attention of

---

* DeepMind recently beat world champion Lee Sedol at the game of Go, long thought to be too complex for artificial intelligence.

the right clinician in time,' a DeepMind spokesperson told me. 'When this doesn't happen, the consequences for patients can be severe, and even fatal. The aim for STREAMS is to speed up the time to alert nurses and doctors to patients in need down to a few seconds.'

The hospital didn't approach the patients for explicit consent to use their data for STREAMS, reasoning that it was part of the clinical care they were receiving in the hospital. However, critical coverage of the Royal Free/Google DeepMind collaboration in *New Scientist* magazine resulted in a wider discussion of the ethical implications of sharing NHS patient data with a private company. Some commentators asked why data from such a large pool of hospital patients was included in a system designed to detect a single condition.

I put this to the DeepMind spokesperson, who responded, 'AKI can affect patients for a wide variety of reasons, including patients who develop AKI as a consequence of another procedure, such as a hip replacement, or because of another medical condition, such as pneumonia or sepsis. That means it's very difficult to predict exactly which patients will develop AKI.'

In the UK, the project of establishing a system gathering all NHS data into one database remains unresolved, so harnessing medical records for research still happens on a piecemeal basis. The tension between personal data privacy and the research potential of sharing health data continues, and will continue until patients can be convinced of the benefits of such research, and the trustworthiness of the systems in place to use their data.

Google DeepMind's response to criticism over the Royal Free Hospital collaboration included a new technological initiative, Verifiable Data Audit, which would ultimately let people know who had accessed their data, when and for what purpose. That type of digital infrastructure could usefully be applied beyond healthcare. 'We plan to make the infrastructure powering this freely available as open source,' the Google DeepMind spokesperson told me, 'so any organisation in the world would be able implement their own version if they wanted to.'

The Royal Free London NHS Foundation Trust is one of 12 trusts selected by the UK government in 2016 as Global Digital Exemplars, to lead the way in harnessing data for better healthcare (and more efficient delivery of increasingly cash-tight services). So no doubt they were motivated by the desire to pioneer new systems that could provide faster, better, cheaper care to their patients.

Another of the Global Digital Exemplars, Salford Royal NHS Foundations Trust, is experimenting with wearable devices that measure patients' activity, sleep patterns and vital signs. By combining some of this data with the patients' electronic medical records, they hope to improve the efficacy of medication and help patients better manage their own treatments. It also fits perfectly within the UK government's Personalised Health And Care 2020 framework, which is all about 'Using data and technology to transform outcomes for patients and citizens', partly through a more efficient system and better medical care, but also through preventing illness by promoting healthier lifestyles.

So your own wearable device that helps you exercise more, sleep better, and so on, becomes joined up with your medical records, and the doctor who may be the gatekeeper for your access to future medical treatment can read the data. And just as smokers and overweight patients today may find themselves sent to the back of the treatment queue, patients who prefer eating a fried breakfast to jogging may wait longer for an operation than their fitness-fanatic neighbour.

The promise of big data feeding research and improving medical care for all of us is immense, and only just beginning to be explored. But it does not take place in socially, politically or morally neutral territory.

Rules and laws

The landscape of law and regulation that governs what may be done with our data is also changing fast.

The UK law replacing DRIPA (as described in Chapter 8) was passed by parliament in 2016. The Investigatory Powers Bill, or IP Bill, delivered sweeping legal powers to collect, retain and access mass datasets, including the kind of metadata that shows what numbers you called (and when, and for how long) and what websites you visited. In fact, it mandates that telephone companies and internet service providers (ISPs) must retain this data for a minimum of one year. Authorities permitted to access this data include not only police and security services, but local authorities, the NHS, the Department for Transport and the Food Standards Agency, to name a few. In some cases, the only authorisation required is sign-off by a senior official within the organisation.

So a senior official in your local council could sign off permission for your mobile phone records, for example, to be examined on request by whoever is responsible for monitoring dog waste offences. If you think this is exaggerated, read what Glasgow's Future City website has to say about people who let their dogs poop outside the authorised areas, and don't clean up.*

Two UK MPs took a case to the European Court of Justice (ECJ), asking for a ruling that the provisions of the IP Bill were illegal under EU law. In December 2016, the ECJ delivered a ruling that 'EU law precludes national legislation that prescribes general and indiscriminate retention of data'

---

* 'Approximately 10 tonnes of faeces is produced by dogs in Glasgow every day. This is clearly unacceptable and one which threatens the health of local community particularly young children. [sic] ... Fixed penalty notices of £40 are issued under the Dog Fouling (Scotland) Act 2003 to offenders who let their dogs foul and do not immediately remove the excrement appropriately. This increases to £60 if not paid within 28 days.'

It's hard to know how to reduce the amount of faeces dogs produce without limiting their food intake or employing some kind of temporary, cork-based intervention. Anyway, you can look at Dog Fouling Complaints (DFC) data on a map at FutureCity.Glasgow.gov.uk.

and that 'the retained data, taken as a whole, is liable to allow very precise conclusions to be drawn concerning the private lives of the persons whose data has been retained.'

It also notes, however, that exceptions to the assumption of privacy may apply for such purposes as national security or fighting serious crime. It is the indiscriminate and wholesale retention of data, with weak or no limits on its access and use, that makes the IP Bill unacceptable to the ECJ.

Although the UK will be leaving the EU, there are signs that the UK government is seeking to quietly revise the IP Bill, or to introduce technological fixes that will render it compatible with EU law.

Another new element in the legal landscape is the EU General Data Protection Regulation (GDPR), which will apply throughout the European Union from 2018. This introduces much tighter limits on what personal data may be collected and retained, and how it may be used.

The GDPR definition of personal data includes IP addresses, as they can easily be linked to individuals or families. It specifies eight rights that each individual now has over their data, including the right to access data about yourself, to have it rectified or erased if incorrect or no longer relevant, and rights in relation to automated decision making or profiling.

The profiling section says that any organisation using AI or other data analysis to make predictions about an individual must put safeguards in place. They must prevent discriminatory effects, minimise the risk of errors, enable error correction, and be fair and transparent about the logic involved and the significance inferred. Areas of an individual's life included in this part of the regulation include performance at work, health, behaviour, personal preferences, location and movements. It's easy to see that this could apply to targeted advertising, workplace monitoring and the health service initiatives described above, and it will be interesting to see how this very general regulation plays out in practice. One loophole is that decisions based on explicit consent, or those necessary to enter into a contract, are not covered.

Again, although the UK will be leaving the EU, it will conform to the GDPR – not doing so would make it impossible to trade or collaborate with most of Europe.

Meanwhile in the United States, the kind of decision-making algorithms used in the criminal justice system, as described in chapters 6 and 9, have become the subject of both public controversy and legal challenge.

A well-researched article by Julia Angwin and others, published in ProPublica[*] in 2016, analysed the role of risk-analysis algorithms in sentencing. They quote then-US Attorney General Eric Holder's warning: 'I am concerned that ... they may exacerbate unwarranted and unjust disparities that are already far too common in our criminal justice system and in our society.'

Then, through meticulous statistical analysis of actual cases, and of which persons did go on to reoffend, the journalists show that one of the most popular systems, COMPAS, designed by a company called Northpointe, did indeed perpetuate the justice system's tendency towards racial bias. Specifically, it was more likely to wrongly assign a high risk of reoffending to black defendants and a low-risk score to white defendants.

There is no suggestion that the system used includes any explicit reference to a defendant's race, or any deliberate proxy to introduce race as a measure. Northpointe objected to ProPublica's analysis and conclusions. But looking at outcomes does reveal that black defendants are more likely to be wrongly assigned high-risk scores both for reoffending and for violent reoffending. White reoffenders are more likely than black reoffenders to have been given a low-risk score. The benefit of the doubt is not being shared around equally. Basing algorithmic decisions on data from the past, where the past is riddled with inequality, very easily perpetuates the

---

[*] https://www.propublica.org/article/machine-bias-risk-assessments-in-criminal-sentencing

very patterns that you were hoping to escape by introducing the objectivity of a machine into your system.

Following this public debate, one of the convicted criminals whose case was described in the ProPublica article, Eric Loomis, appealed against the use of COMPAS in deciding his sentence. As Marion Oswald and Jamie Grace describe in their 2017 article, 'Norman Stanley Fletcher and the Case of the Proprietary Algorithmic Risk Assessment':

> *Loomis asserted that the court's consideration of a COMPAS risk assessment at sentencing violated his right to due process. In particular he argued that because the tool was based on group data, it violated his right to an individualised sentence (the tool was designed to predict group behaviour); and the proprietary nature of the tool prevented defendants from challenging the assessment's scientific validity.*

The Supreme Court of Wisconsin agreed with his point that the algorithm's predictions were of a general, statistical nature, not a specific judgment about what Loomis, personally, would or would not go on to do. The court commented in the ruling that 'risk scores are intended to predict the general likelihood that those with a similar history of offending are either less likely or more likely to commit another crime following release from custody. However, the COMPAS risk assessment does not predict the specific likelihood that an individual offender will reoffend. Instead, it provides a prediction based on a comparison of information about the individual to a similar data group.'

The court also noted that COMPAS was not developed for sentencing, that there were concerns about ethnic bias and that the proprietary nature of the algorithm prevents disclosure of how scores are assigned.

Nevertheless, the Supreme Court of Wisconsin ruled in 2017 that, 'Although Loomis cannot review and challenge how the COMPAS algorithm calculates risk, he can at least review and challenge the resulting risk scores set forth in the report.' It therefore dismissed his claim, though it did also state that such risk scores should not be used alone to determine a sentence, and that courts should exercise discretion when applying a score to an individual defendant.

The regulation climate in the United States remains very different from that in Europe, not only because of the US Constitution's presumptions in favour of free speech, freedom of association and conscience, and freedom from general searches, but also because it relies more on redress against misuse of data, rather than statutory control of collection and retention.

In early 2017, the Trump administration began measures to roll back regulations preventing ISPs from selling on detailed data about the online activities of their customers. As the Electronic Frontier Foundation pointed out, while you can avoid using Google's services to limit their collection of data about you, it is much harder to be selective about which part of your internet activity passes through your ISP.

Political objections to this official sanctioning of wholesale data collection and sale may at least have alerted the US public to how little privacy they could have every time they go online, and provoke either political resistance or the desire to find technological solutions that defend their privacy.

On a more positive note, Oakland, California remains a beacon of how active citizens can hold to account those collecting and using their data. The ad hoc Privacy Committee described in Chapter 8 is now a statutory Privacy Advisory Commission. At its first meeting in July 2016, Brian Hofer was elected chair.

Its task is to draft laws that will govern the acquisition and use of future surveillance technologies in Oakland, and define a process through which public discussion will always happen before new technology is bought and used.

At the time of writing (early 2017), Oakland Privacy is drawing up a standard questionnaire for use when assessing new proposed technologies. Draft questions include, 'What is the specific problem this equipment or use will resolve?',

'Does the technology collect and retain information about individuals not suspected of wrongdoing? If so, how could such information impact their right to privacy?' and 'Could the technology be used on groups, public gatherings or crowds and thus have an effect on First Amendment activities such as protests? If so, what safeguards are in place to limit this?'

Oakland Privacy has already drawn up a statute controlling police use of the 'Stingray'-type IMSI catchers, or cell site simulators, described in Chapter 8. Brian Hofer wrote to Oakland City council, describing the policy the Commission had adopted in early 2017 as a gold standard. Content interception is prohibited, and each use requires a warrant and subsequent entry in an annual report. 'There is no policy regarding cell site simulators in existence even remotely close to Oakland's,' says Hofer, 'as to the narrowness of allowable use, oversight and transparent reporting.'

Now other organisations turn to Oakland Privacy and the ACLU for help in drawing up their own laws and regulations. Santa Clara, Palo Alto, Berkeley and even the regional transport authority, BART, have used Oakland's Ordinance as a template, and asked for help in adapting it to their own needs. Brian is kept busy speaking at meetings and offering advice.

'People are excited that a small volunteer group has been able to achieve so much success,' he told me in 2016. 'A large part of that success is due to coalition building. By involving civil liberties organisations, labour, social justice causes and all sorts of regular folks, we are able to show that a great many people demand reform in this area.'

The Privacy Advisory Commission's work is only just beginning. But Oakland is an inspiring story of what people can achieve when they demand to have control and oversight of the technology being used to gather data about them.

Was it big data wot won it?

Although technology and its practical applications have pushed on faster than ever since I completed the first edition of this book, I think that the social and political discussions around big data are equally important.

In Chapter 7, I discussed the role of data in the UK General Election of 2015. Since then, the UK has voted to leave the European Union, the USA has elected President Donald Trump and the UK will have held another General Election before this edition goes on sale.

Some news reports attributed both the Leave vote and Trump's victory to the use of big data techniques, and in particular to the work of one company, Cambridge Analytica.

Cambridge Analytica uses what it calls 'Behavioural Microtargeting' to 'deliver the right messages to the right individuals in meaningful ways online'. It promises to combine behavioural psychology with data analysis to target individuals with the most effective messages to get them to do what you want – to vote for your candidate, for example.

Of course, companies like Cambridge Analytica want you to believe that their software can unlock the secret desires of millions of people and bend them to your will. That's their business model: promising to give you the edge over political opponents or commercial rivals.

But as you saw in Chapter 7, none of this is particularly new.

Barack Obama's election campaigns harnessed social media and detailed use of data about potential voters, which at the time was hailed as evidence of his modern, enlightened approach to politics. This included collecting the consumption and behaviour patterns of the intended audience and showing them a customised message, instead of the same message everyone else was seeing.

In *Data and Democracy*, published by O'Reilly Media in 2016, Daniel Scarvalone of Bully Pulpit Interactive (BPI) describes how it uses 'Experimentally-Informed Programs' (EIPs) to measure which adverts are having more effect, and adjust as they go.

None of this is very different from the way companies try to sell you their products. Online advertising is not just targeted, using what is known about you, but also tested on you for effectiveness. Even the UK government's Behavioural Insights unit tested eight different online messages designed to 'nudge' people to register as willing organ donors. They measured which one resulted in the most new registrations, and that's the one you'll read when you tax your vehicle online.

Nor is Scarvalone bashful about using commercial platforms and companies to work more effectively. 'No one has built a person-based advertising platform that rivals Facebook's, [or] a buying platform with breadth and reach that rivals Google's Doubleclick,' he says in the same publication, also noting that 'an overwhelming majority of Democratic campaigns use ... Acxiom, Experian or Infogroup to supplement their voter files with consumer records.'

On one level, using big data to target persuadable voters is nothing more than a change of medium. Politicians have always sought to win enough voters to help them win elections. Since democracy came to mean mass suffrage, that means winning over masses of ordinary voters like you and me. In the past, they might have used posters, radio or TV campaigns, public meetings or door-to-door canvassing. Now they're moving online.

Turning to new technology reflects, in great part, the loss of these other channels of communication. The era of mass membership political parties has passed, along with other forums for politicians to engage with the general public in active debate. How else can they find out what we care about, what gets us angry, inspired or afraid, except by what we Tweet, Like or Share on the internet? How else can they seek to win us to their party or their cause, except through the same social media?

When Trump used these tools and strategies, however, some media coverage took on a different tone. The UK's *Observer* newspaper (sister publication to the left-leaning *Guardian*) reported LSE academics demanding urgent reform of UK electoral law to protect our democracy from big money using big data to manipulate voters (Cambridge Analytica's majority shareholder is Robert Mercer, the hedge fund billionaire who backed Donald Trump).

But the problem lies not in the digital medium, but in the content of the messages.

Seeing politics as a marketing exercise predates the technology. Focus groups and surveys have been used to shape the campaigns – and even the policies – of major political parties for many years. Without strong ideas to distinguish them from their rivals, all parties rely more and more on finding out what preys on the minds of potential voters, and reflecting those preoccupations, hopes and fears back to them in the most persuasive form.

Profiling the electorate drives this reduction of politics to 'whatever jerks your chain' still further. Instead of finding policies that address the shared interests of large sections of the population, you may be able to win elections by identifying the hot issues of several distinct subsections, and tailoring campaigns to them specifically. For example, to working-class voters in stagnating regions you can promise 'British Jobs For British Workers', while bigging up your anti-racist credentials for the cosmopolitan urbanites.

The deep, underlying problem here is not the technology that helps campaigners guess which messages will drive you most effectively to turn up and vote for them. The worst aspect of the data-driven campaigns is that they see us not as reasonable voters who may be convinced by a stronger argument, but as participants in a behavioural experiment. Just as the Behavioural Insights unit started with a few familiar psychological principles, and then counted the clickthroughs on each variant of the donor registration appeal, data-based election campaigns can test which adverts and appeals generate the correct action. It's a stimulus–response model, concerned less with underlying motivations than with outcomes.

Again, this loss of belief in the voter as a fundamentally rational person, capable of harnessing both passion and reason when deciding how to vote, precedes the technology. Once you lose faith in being able to win masses of voters to your ideas through argument, using technology to find the most effective stimulus for the desired response is an inviting solution.

This is why the idea that data-driven, targeted advertising won the election for Trump, and the Referendum for Leave, is more than an eccentric conspiracy theory. It is the mirror image of the thing it most fears. Only an electorate incapable of rational thought, impervious to common sense, context and conflicting arguments could be swayed against its will by online adverts, however well-aimed. Begging the question, why is this nightmare vision more inviting to so many than the alternative scenario: that the electorate was simply not convinced by the arguments for Clinton, or for Remain?

On some level, perhaps it is easier to blame defeat on a conspiracy of right-wing funders and mysteriously powerful technology than to accept that convincing others to share your ideas is harder work than you realised.

In the week I wrote this updated chapter, I spent an evening with around 100 'Skeptics in the Pub'* in the Midlands city of Nottingham, talking about big data.

One person asked if I thought big data could ever hoodwink an entire population into something like going to war. I said I didn't think so, for the same reasons I have outlined above. Many of those present had come along because of targeted Facebook ads, and when I asked whether they felt they'd been brainwashed by data-driven social media adverts into wasting an evening, they laughed. Some may have privately felt that, as self-identified Skeptics, they were less easily fooled than gullible others, but I think they got the general point.

Another person suggested that big data was a new moral panic, and that we were too quick to see the risks and too slow to see the life-enhancing potential.

---

* A nation-wide monthly meeting, at which people can discuss skepticism around various topics of interest and debate.

I think there is some truth in that. And I confess that I am always raising the questions of privacy, of profiling and of reducing human beings to what about them we can easily measure and collect as data. I spend less time sharing my excitement about what people are achieving with data. That's not because I see big data as something we should resist; I argue that its potential is under-explored. There is nothing inherently negative in the technology.

Where big data appears to be taking over areas of human life where it doesn't belong (in my opinion), that's not down to a sinister AI trying to take over the world. It's because of our own tendency to trust technology (that we often don't understand), rather than ourselves and each other.

Who knows where the engine of big data will have taken us by the time you read these words? It's impossible to predict. But it's not impossible to keep human hands and minds on the steering wheel, and human eyes on the roadmap, as we hurtle into the future.

It's going to be a thrilling ride. Don't be a passenger.

Appendix

Keeping your data private

There is no utterly secure way to keep your data completely private, except to not collect it, or allow anybody else to collect it. Those who want illicit access to your data are constantly working on ways to get to it, and those who are trying to protect your privacy are constantly working on new tools to lock it away. It's an arms race. So the more precise the tips I offer here, the less likely they'll still be accurate by the time you read them. For up-to-date tips, try the EFF (Electronic Frontier Foundation), Privacy International, and other privacy activist groups. What I offer here is quite general, in the hope it will remain useful for the life of the book.

The first thing is to decide what you want to keep private, and from whom. Or, to turn it around, what would you be happy to share, and with whom? If you're an activist in a repressive state, or an investigative journalist, you'll need more sophisticated techniques than the average person who just wants to be more in control of who knows what.

The second thing is simply to take control of what you choose to share. Most of us won't be the target of elaborate hacking, or of government surveillance, but we will unthinkingly give up private information simply because somebody asks. This may be used for marketing, to add to a database that could then be sold to law enforcement agencies, charities or political parties, or stolen by fraudsters.

So, before you type in your personal details to that online survey, or post personal information to social media, stop and ask yourself a few questions. Who will have access to this data in future? What am I getting out of this? Will it be worth the potential future cost of adding to the available data about me?

If you want free stuff without adding your email to yet another list, why not set up a separate email account specially for mailing lists? You could be super-cheeky and call it 'spam@whatever.com'. That should quarantine your spam, but it's quite likely that it will be linked to your other online identities eventually.

And remember that you may be sharing other people's data. Directly, if you tag a photograph of them in the pub when they should be at work. Or indirectly, when you allow an organisation access to your Friends list, or give your family home address to a database.

Passwords

As a lazy person, I know the temptation to use one memorable password for everything from banking to ordering pants online. Don't. Anybody who gets that password by hacking the pants retailer, for example, can now access all your business. Choose passwords that are different, and difficult, and use a password manager program that can store them securely for you. KeePassX is recommended by Edward Snowden, and I assume he knows what he's talking about. Or you could write your passwords on a bit of paper and hide that somewhere private, disguised as a shopping list or something.

Many services offer two-factor authentication, a password backed up with a text message to your cellphone, for example. That makes it much harder for somebody to access your social media or email accounts, even if they get hold of your password.

Security questions

Could somebody who looks at your Facebook page find your date of birth, mother's maiden name, name of your pets and children? Then they're not very secure as security answers, or as passwords. Choose less obvious security questions, and don't use the same ones all the time.

And remember, you don't always have to be truthful. Who's to say your mother's maiden name wasn't 'ofAragon' or your first pet wasn't 'Smaug the Dragon'? Some people use several, fictional, profiles online, each with their own date and place of birth, family names, etc. If you're a budding novelist with a good memory, that could be fun.

Encryption

There are readily available options to encrypt much of your personal data, on your phone and on your computer hard drive. That means anybody who steals the hardware won't be able easily to read the contents.

You can also use apps and services that offer end-to-end encryption, so what you share is secure both inside the device and en route to the recipient. The app Signal and Apple's own messaging service currently offer this. This could change if the UK Government ignores all advice and insists on having backdoors built into encryption.

Browsing the internet

Using an Ad Blocker will protect you from some hacking attacks, as well as from unwanted targeted ads. If you're annoyed by the targeting process you can often turn it off. Google, for example, allow you to opt out of 'ads based on your preferences' and receive only generally targeted ads, based on your location rather than your browsing history.

You could use the EFF's browser plugin HTTPS Everywhere, which will try to encrypt all exchanges with all websites in the same way that banking and other financial interactions are already protected. Options called 'private browsing' aren't all that private, though they will limit what information the websites you visit can access about you. My flights to America to research this book were cheaper because I used private browsing to compare prices.

For more private browsing, use the TOR browser, which is free. It is slower, and some sites may not co-operate with it, but it does make you much more anonymous online.

Social media

Most social media sites offer a range of privacy settings, so you can have some control over who sees what. However, the best protection is to think about what you post, and talk to your friends about what they post. It's much harder to remove something than to delete before you post.

The fact most of them are provided free should alert you to the fact that Facebook, Twitter and their ilk regard you and your data as the product. They collect your posts, along with such information as where you are, what kind of machine you are using, and other aspects of your life that let them build up a profile of who you are.

If you don't like this, you are free to opt out, though this can make your social life more difficult. You can obfuscate the data about you by posting misleading information, such as a wrong date of birth or relationship status. Or you may accept that it's the price of a free service. But make an active decision – don't just give up your privacy without thinking.

Cellphones

As we've seen, cellphones are the most sophisticated, near-universal tracking device ever carried by the majority of human beings. Smartphones have microphones, cameras, location recorders, movement detectors, and a whole range of communication information. Even when turned off, they can reveal your whereabouts and be vulnerable to interception.

The only way to completely prevent this is to put your phone inside a Faraday Bag that prevents telephone, WiFi and Bluetooth signals from reaching the phone. You can buy these bags online. Naturally, this means it can't be used as a phone when it's inside the bag.

A less extreme approach is to think about what apps you have on your phone, and what you allow them to do. Most smartphones allow you to control whether apps can use Location Services, for example. So, though the FBI or MI5 might be able to hack into your phone, at least every app you've ever downloaded won't have a record of your daily movements.

If you allow your smartphone to connect to other people's WiFi connections, perhaps because you're abroad and don't want to pay roaming charges, you will give up certain information in exchange. Free WiFi may ask for your email address. Even if it doesn't, it can get certain information such as the IP addresses to which you most often connect and, by inference, the areas where you live and work. In many cases, it can collect this data even if your phone doesn't go on to connect with their network. If you're not happy about this, turn off your phone's WiFi option when out and about.

Or you could leave the phone behind sometimes. Radical, I know.

If you don't like the idea of your every interaction, movement and activity being tracked, recorded and potentially examined without your permission, there are two more things I'd recommend.

One is that you get involved in one of the campaigns for more controls over what data can be collected, stored and used, and by whom and for what purposes. Privacy International are just one of the organisations working hard in this area. Some of the discussions are very technical, but many are around points of principle. Should we know when our data is collected and stored? Who has oversight of the security services using surveillance of our communications? Is a contract so long that most of us don't read it a fair way to get our consent? I can't tell you what you should think on these and other issues, but I can say that it's important you get involved in the debate, and tell your elected representatives what you think.

The other thing you can do is remember that you do have a private life, and that it doesn't have to be lived digitally. Leave the smartphone at home, have a conversation face to face, send a letter, write your thoughts in a paper diary. If you go for a romantic walk in the forest and don't post it to Instagram, it still happened.

It's yours, it's private. And that's important.

Acknowledgements

This book has been in gestation for about five years. During that time I have chaired and spoken at many public discussions of big data, made a radio documentary about it for the BBC, read a pile of books, and talked to hundreds of people both on and off the record. Thanks to all those people. You are too numerous to name. If you suspect that you've helped me develop the ideas here, you are almost certainly right, including (or especially?) those who disagreed with me. Talking through ideas is the best way I know to have better ideas.

Special thanks are due to Rob Lyons who has acted as something between a Ph.D. supervisor and a joke-writer. Also to Martin Rosenbaum, who made the Radio 4 documentary *Data, Data Everywhere* with me, and to Michael Blastland, who has helped me clarify my thoughts on many train journeys between Clapham Junction and Brighton.

My mathematical and statistical understanding increased about a million per cent thanks to David Spiegelhalter, Scott Keir, Hetan Shah, Jennifer Rogers, the Royal Statistical Society and the Open University Mathematics and Statistics department, among many others.

Big data thoughts were given extra dimensions by talking to Sandy Starr, Dr Norman Lewis, Professor Gary Marcus, Dr Marion Oswald, Dr Tiffany Jenkins, Professor Tim Cole, Dr Nick Hawes, Dr Ellie Lee, Dr Brett Lempereur, Dr Philip Hammond, Professor David Chandler, Josie Appleton of the Manifesto Club, Matt Cagle of ACLU, Fran Bennett of Mastodon C, and Steve King of Black Swan.

Support, advice, introductions and sometimes unwitting contributions came from Matt Parker, Dr Helen Pilcher, Claire Fox, Paul Thomas, Sandra Lawrence, Professor Chris Lintott, James Barrett, Dr Hannah Fry, Dr Andrew Pontzen,

Dr Tom Whyntie, Hilary Salt, Tom Ziessen, Anne Gammon, Daniel Tyrrell, Jonathan Brunert and Team Bad Sauna.

Gareth Roberts, Matt Pritchard and Terence Eden contributed extra material.

Thanks of course to my editors Jim Martin and Anna MacDiarmid, Nick Ascroft and the rest of the team at Bloomsbury Sigma, and to Blacks for espresso martinis.

Finally, thanks to my family, friends and flatmates who have supported me through the whole process, tolerated my unreasonable demands, made sure I ate some vegetables, and generally contributed the kind of essential human support that cannot be quantified on any database. Without you, there would barely be me, let alone a book. Thank you all.

Index